Matthias Hofmann

Data Management for Natural Scientists

Also of Interest

Empathic Entrepreneurial Engineering
The Missing Ingredient
David Fernandez Rivas, 2022
ISBN 978-3-11-074662-4, e-ISBN 978-3-11-074682-2

Big Data Management
Data Governance Principles for Big Data Analytics
Peter Ghavami, 2021
ISBN 978-3-11-066291-7, e-ISBN 978-3-11-066406-5

Data Science in Chemistry
Artificial Intelligence, Big Data, Chemometrics and Quantum Computing with Jupyter
Thorsten Gressling, 2021
ISBN 978-3-11-062939-2, e-ISBN 978-3-11-062945-3

Risk Management and Education
Thierry Meyer, Genserik Reniers and Valerio Cozzani, 2019
ISBN 978-3-11-034456-1, e-ISBN 978-3-11-034457-8

Matthias Hofmann

Data Management for Natural Scientists

A Practical Guide to Data Extraction and Storage Using Python

DE GRUYTER

Author
Matthias Hofmann
Peter-und-Paul-Weg 1
84558 Tyrlaching
Germany
matthias.j.hofmann@gmx.de

ISBN 978-3-11-078840-2
e-ISBN (PDF) 978-3-11-078843-3
e-ISBN (EPUB) 978-3-11-078853-2

Library of Congress Control Number: 2022949062

Bibliographic information published by the Deutsche Nationalbibliothek
The Deutsche Nationalbibliothek lists this publication in the Deutsche Nationalbibliografie; detailed bibliographic data are available on the Internet at http://dnb.dnb.de.

Cover image: top: Gettyimages / phototechno; bottom: Gettyimages / Mironov Konstantin
Typesetting: VTeX UAB, Lithuania
Printing and binding: CPI books GmbH, Leck

www.degruyter.com

Acknowledgment

Writing this book would not have been possible without the help and support of the people I had the pleasure to work with along my journey. In particular, I want to thank Prof. Dr. Torben Gädt and Dr. Tobias Lange from TU Munich for critically reviewing the manuscript and their thoughts on the essential steps across the herein described data processing workflow.

Furthermore, I would like to thank the team at De Gruyter for giving me the opportunity to write this book as it turned out with the intention of serving practitioners in the natural sciences to advance their data-handling, analysis and visualisation capabilities.

Finally and endlessly, my gratitude goes to Katharina. After the initial *Why another book?*, you offered the continuous support and indulgence needed during the time of writing. Thank you.

Mission

The book is intended to provide a "hands-on" guide to lower-level data processing routines for data generated in the context of the natural sciences, such as chemistry, material science and the like. As in many practical use cases, raw data files obtained from an experimental device will serve as the starting point for the considerations herein. In most of the typical scenarios, these raw data files hold complex information and are best read by proprietary software provided by the respective manufacturer. According to the *data pyramid* scheme, *data—information—knowledge—wisdom*, the first step is to extract the *data* from the raw data file. Next, the extraction of *information*, i. e., the characteristic parameters from the collected *data*, will be demonstrated via the use of easily accessible tools. In particular, the focus is in on the high-level programming language Python for the critical transformation step from *data* to *information*.

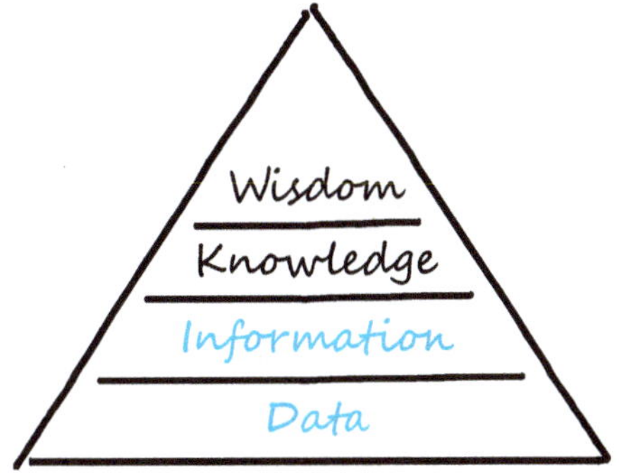

https://doi.org/10.1515/9783110788433-201

Furthermore, the multiple options for storing the extracted data and information–-from an easily structured file browser system across a local database to more advanced database systems—are treated, but the latter will not be the focus of the book. Finally, options for visualisation of the processed experimental data will be presented and possible methods for more detailed analyses will be explained. The process will be walked through on the basis of a simplified real-life example.

Audience

The book is intended to address graduate students of the natural sciences (probably excluding such "number-crunching" disciplines as genomics, etc., that use more advanced techniques) who seek to improve their data handling skills and capabilities for long-term accessibility of their experimental results. Likewise, the book's potential audience includes university chairs who want to ensure a sustainable long-term use of experimentally collected data across multiple "generations" of students. Another possible target audience could be smaller or midsize companies in the described fields not able or willing to employ a full-time developer for handling advanced database environments and scientific staff with a strong background in data management. In this sense, the book's goal is to promote a mutual understanding of the needs of both scientists and data engineers.

Assumptions

This book does not require in-depth previous knowledge on scripting in Python or data management, but familiarity with basic concepts such as the creation of folders and files is assumed. Some previous experience using scripting languages is certainly helpful but not mandatory for understanding the key ideas of this book and following the provided code snippets.

The examples presented herein were created using Python 3 on a Windows platform.

This book is not …

This book is not to be understood as a full-stack reference for all of the *packages* and functions used to organize the data related to the exemplary scientific question. Hence, there will be no in-depth explanations of *arguments* and *keyword arguments*, i. e., the arguments passed to a Python function. This is especially the case for those arguments that are not used in the scope of the presented examples. At this point, reference is made to the rich and current online documentation of the relevant packages.

For those sites, the documentation matching your installed version is readily available.[1,2,3] Furthermore, this book does not claim to achieve programming excellence or syntactic perfection in each line of code. Rather, it condenses the lessons obtained over the last few years as I worked with Python in a scientific context on a daily basis. In short, this is a book by one *practitioner* for the would-be practitioner.

Contents of this book

Chapter 1: Presenting the challenge

Experimental data comes in many forms. In order to ensure long-term sustainable access to experimental data generated and obtained with some effort, the following challenges, herein formulated as questions, have to be addressed by a anyone planning a well-defined data management system for natural scientific data:

- How to make the results accessible to yourself, your colleagues, your institute and/or your employer?
- How to make best use of the data? This includes drawing meaningful conclusions based on experimental evidence in order to allow for data-driven decision making.
- How to deal with new questions once the data collection has been completed?
- How to reuse experimental results?
- How to share *data* and *information*?

Chapter 2: Python quick start

Herein, the very basis of getting your Python installation up and running are addressed. Additionally, the key concepts of *scripting* used in the remainder of the book are introduced.

- How to install Python via distributions (such as WinPython or Anaconda)?
- Why is Python used herein?
- The *Zen of Python*.
- Key data structures.
- Key packages.

Essential examples of code are shown, and the basics of prototyping/scripting in the integrated development environment (IDE) *spyder* are explained.

1 https://pandas.pydata.org/docs/

2 https://seaborn.pydata.org/api.html

3 https://matplotlib.org/stable/api/index

Chapter 3: The steps of data processing

Probably, *the* challenge in a scientific, or likely in any other, field is to move up the *data pyramid* as defined within this chapter. According to this concept, *information* is defined in terms of *data*, *knowledge* in terms of *information* and *wisdom* in terms of *knowledge*. Critical for understanding is the notion that we cannot climb to the upper regions of this pyramid without having transversed the lower stages. In essence: There is no *wisdom* without *data*.

This problem is also linked to how we typically obtain results from experimental devices. Most of the time, they merely provide *data*, i. e., the lowest level of the *data pyramid*, in a more or less accessible way. The effort to master this *data*, i. e., to visualize the results, feels like a huge step. In reality, however, we haven't yet moved up to the next step of the pyramid. We just made the first step.

Developing a sound understanding of how to access the bottom levels according to this schema, *data* and *information*, will be the focus of the major part of this book.

Chapter 4: From experimental files to data

Getting access to the *data* part of an experimentally obtained results file is key to all of the downstream steps in data processing. So, let's start our Python scripting endeavour at the beginning and tackle the problem at the source. The following topics will be addressed:

- How to extract and separate *data*, *metadata* and *information* from an experimental results file?
- How can *regular expressions* assist us in this task?

Chapter 5: From data to information

This can be justifiably regarded as the "fun part" for those experts in the respective fields having the storied *domain knowledge*. Long-standing hunches can be "translated" into quantifiable parameters based on experimental data. A four-step approach is suggested for extracting *parameters* (relevant and/or characteristic numbers), i. e., *information*, from the now available *data* (pure numbers without additional context):

- visualize collected *data*,
- extract parameters generally accepted in the field (state-of-the-art parameters),
- extract parameters that *might* be interesting for the purpose of a particular analysis but are not regularly considered in the field, and
- compile a set of parameters corresponding to a sample.

To illustrate the idea, let's consider some exemplary results from thermogravimetric analysis (TGA). In this method, the mass of a sample is measured upon heating. "Steps" in the derived relative mass versus temperature plots are associated with characteristic chemical decomposition processes. From those, we extract a temperature, *T1*, at which the maximum change in mass upon increasing temperature occurs. For the sake of the example, this is to be understood a the *standard parameter*. Furthermore, we extract the remaining weight at a specific temperature, *w2*, as a *custom parameter*. This value might be relevant, e. g., for processual reasons such as ensuring minimum contents of a residual component given the defined process conditions restricted by the setup.

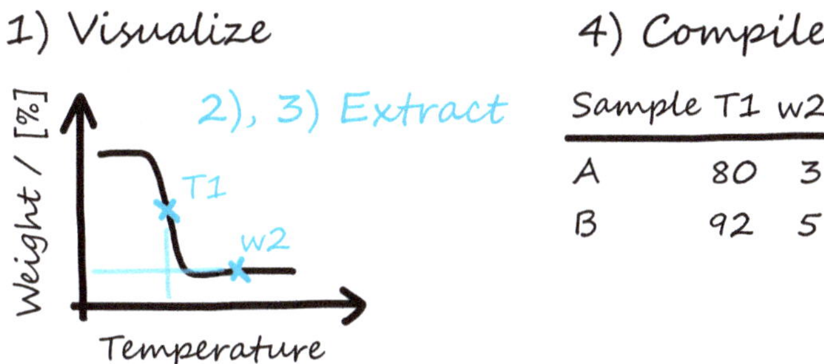

In the practical example presented later in this book, parameters are extracted using a Python script.

Chapter 6: Where to put data and information

Storing both *data* and *information* in one or the other way is a critical step to ensure long-term accessibility and thereby long-term use. Depending on the volume of data and possible restrictions in your environment (such as access and admin rights), there are approaches spanning several levels of complexity:

- Basic: using a folder structure populated via scripts using strict naming conventions. Files should not be opened and modified manually. A negative aspect of relying on this method is the requirement to maintain strict discipline.
- Intermediate: using local database files, such as *SQlite*, coming with Python or *Microsoft® Access®*.
- Advanced: hosted databases such as *MySQL* or *PostgreSQL*.

The general idea is merely to establish a "space", where *data* and *information* are provided in a readily searchable, findable and accessible way. In this chapter, two examples are shown in great detail:

- Saving to an organized folder structure via a Python script.
- Saving to a *SQlite* database via `SQLAlchemy` (basic `SQLAlchemy` intro).

Chapter 7: How to visualize data and information

Once stored in the folder structure or database, there are multiple ways to visualize spanning diverse levels of user friendliness and complexity:

- Python script plus visualization libraries such as `matplotlib` and/or `seaborn`.
- Drag-and-drop solutions and dedicated software (Microsoft® Power BI Desktop®, Tableau, Origin).
- Custom Graphical User Interfaces (GUIs). Due to the limited scope of this book, this topic will not be covered in detail.

Chapter 8: Responding to lessons learned/posterior information

In most realistic cases, there's some insight generated in the course of conducting and analysing experiments. Ideally, you extracted the *right* parameters, i. e., pieces of *information* from collected *data*, right from the beginning of a series of experiments or a project.

Another common scenario is the situation in which you find at some midpoint or at the end that other parameters should have been collected—either for your own experiments or for those performed by colleagues in the past.

Having access to the *data*, extraction of further parameters is still possible in hindsight. Herein, a way to add further parameters to an already existing set of data and information is shown using a SQlite database via Python and `SQLAlchemy`.

Returning to the previously introduced TGA example, you might find that the *onset temperature* of the maximum change in observed mass, T3, is a highly relevant parameter. Accordingly, this value can be readily extracted from the *data* and included in the already existing list.

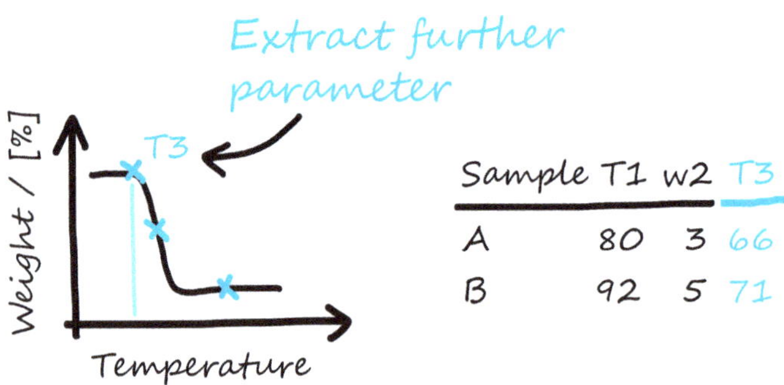

Chapter 9: Where to go from here

So are you still waiting for *knowledge* or even *wisdom* according to the introduced concept of the *data pyramid*? Now, with your *data* and *information* sorted out, there's at least the chance to go there via techniques popularised by the huge field of data science.

The first steps could involve *linear modeling*. But also more complex approaches, such as *neural networks* or other *machine learning* techniques, can be applied (for larger data sets) to extract insights and conclusions from your experiments.

On a smaller scale, i. e., when one has no access to thousands or even more comparable experiments, scientists are typically attracted to linear models due to their simplicity and clear *cause–effect pattern*. For instance, if parameter A is increased, value B also increases (all other things being equal).

A widely neglected pathway in the field of natural sciences is probably rigorous *causal analysis*. Therein, the aim is to draw cause-and-effect relationships between observable variables and results via a previously defined hypothesis represented by a *graph*. In order to shed light on this highly interesting mode of operation, a small example coined to the world of natural sciences is given at the end of this chapter.

Chapter 10: Conclusion

The steps of the data processing routine described in the book are reiterated with a focus on the challenges associated with each of them.

Conventions used in this book

The following typographical conventions are used in this book:

Italic

for new terms, URLs, filenames, file extensions, pathnames and directories.

`Typewriter`

for commands, options, variables, attributes, keys, functions, types, classes, methods and modules.

Small caps

for Structured Query Language (SQL) keywords and queries.

This environment indicates some additional information.

This environment indicates attention to be directed to the described issue.

This environment indicates the option for some exercise based on the mentioned code snippet.

Concerning code snippets

Attribution of the code snippets considered useful for your project is appreciated but not necessarily required. An attribution regularly includes the title, author, publisher and ISBN. For the present case, this could be "*Data Management for Natural Scientists* by Matthias Hofmann, De Gruyter, 2023, ISBN 978-3-11-078840-2".

The exemplary experimental results files and Python scripts shown in the following are available free of charge on
- https://www.degruyter.com/document/isbn/9783110788433/html, and
- https://github.com/mj-hofmann/Data-Management-for-Natural-Scientists.

Microsoft® copyrighted content

Contents

1 Presenting the challenge

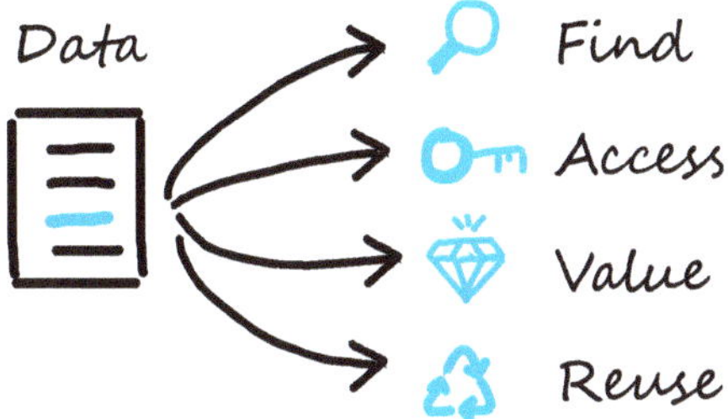

Experimental data comes in many forms. Mostly, we will be dealing with the situation of *data* connected to an individual sample or experiment. This *data* can come in the form of an *individual data point* or a *single value* corresponding to this sample or ranging over *multiple values* to *large sets* of data consisting of tens, hundreds or thousands of points. Depending on the discipline and focus area, there may be large differences in the perception of what *large* means.

As an example for demonstrating the key concepts, I would like to refer to a basic concept for the description of adsorption behaviour in surface chemistry: *surface tension isotherms* of aqueous surfactant[1] solutions. Have no worries since there will be some more introduction to this characteristic in the following, but let's set aside the scientific part for the moment. In this case, *data* on either level can be observed and collected: individual data points (alternatively, values or scalars), multiple data points and larger data sets. The experiment "dissolve a surfactant in water at various concentrations and measure the resulting surface tension at the given temperature" can be described on each of the following levels:

- *Single value* or *individual data point*: Surface tension isotherms are frequently measured for pure surface active agents in water. The molecular mass of this surface active agent, which is a single number, can be used to describe a characteristic property of this chemical, and therefore also the experiment to a certain degree.
- *Multiple values*: The surface-active molecule used in the experiment has a characteristic molecular structure and chemical composition. Both properties can be represented by characteristic strings. The simplified molecular-input line-entry system (SMILES) specifies how atoms are connected one to another inside a molecule. This is achieved via short ASCII-strings.[2] Most molecule editors are

1 *Surfactant* is the shorthand notation for *surface active agent*.

2 ASCII stands for American Standard Code for Information Interchange. It is a common *character encoding* for text data. Character encoding describes a particular procedure for assigning numbers

https://doi.org/10.1515/9783110788433-001

capable of recognizing SMILES and representing the corresponding molecular structures. One of the most heavily studied molecules in the field of surface chemistry is sodium dodecyl sulfate. Its chemical formula is $C_{12}H_{25}NaSO_4$, and its SMILES code is CCCCCCCCCCCCOS(=O)([O-])=O.[Na+]. The surfactant's molecular structure is shown here as just another representation of the molecule.

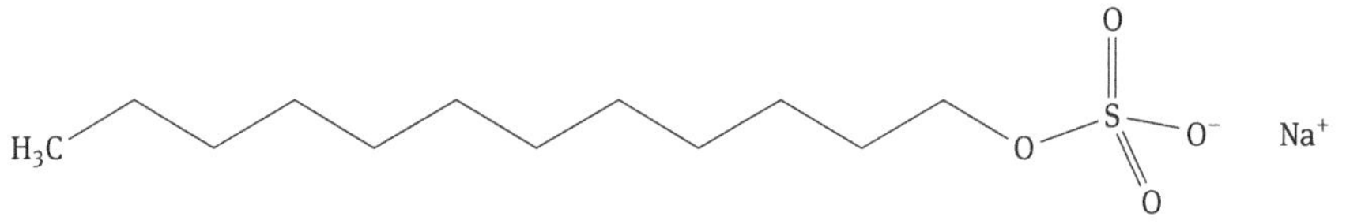

If we want to describe the exemplary surface tension experiment with respect to the molecular structure of the applied chemical, there is more than just a single representation. Here, the structural formula and the SMILES code equally provide us some information about the bonding within the molecule. Other than for the previous *single-value* case, there is more than one correct answer, hence the term *multiple values*.

- *Larger dataset*: Surface-tension isotherms are constructed from ordered lists of surface tension, corresponding concentration and the prevailing temperature. Those triplets are referred to as *tuples*. Depending on your or your lab team's goal, e. g., a surface tension isotherm can consist of only five or tens of tuples instead. If you have dedicated machinery at your disposal, however, surface-tension isotherms consisting of hundred or more tuples are readily available.

A summary of the just mentioned types of data encountered within a surface tension isotherm experiment is given in Table 1.1.

Even from these simplistic considerations, it is obvious that there is a number of things to do "right", but also to do "wrong". Discarding the scientific questions of how to capture quantities of interest and how to do this in the right manner, the focus of this book will be on the following major challenges:

- How to make the results accessible to yourself, your colleagues, your employer or institute or even, more broadly, mankind? Collecting experimental data is one thing. Finding it at a later time—with all the required context—is a completely different story. Finding and accessing data will be an essential part of this book.
- How to make the best use of the data? Drawing meaningful conclusions based on experimental evidence is probably the main task in a scientific context. Certainly, this is also highly relevant for decision making in a data-driven organization in

that can be processed by computers to characters readable by humans. In standard ASCII-encoded data, there are unique values for 128 alphabetic, numeric or special additional characters and control codes.

Table 1.1: Describing a surface tension isotherm experiment.

Type of data	Quantity of interest	Exemplary value
Single value	Molecular mass of pure surfactant	288.372 g, mol^{-1}
Multiple values	Molecular structure and chemical formula	$C_{12}H_{25}NaSO_4$, CCCCCCCCCCCCOS(=O)([O-])=O.[Na+]
Dataset	Experimental tuples of concentration, surface tension and temperature	0.00239 mol L^{-1}, 69.7 mN m^{-1}, 24.6 °C 0.01014 mol L^{-1}, 64.9 mN m^{-1}, 24.6 °C 0.01809 mol L^{-1}, 58.3 mN m^{-1}, 24.8 °C 0.03286 mol L^{-1}, 49.8 mN m^{-1}, 25.1 °C 0.04293 mol L^{-1}, 45.7 mN m^{-1}, 25.2 °C 0.06350 mol L^{-1}, 39.9 mN m^{-1}, 24.7 °C 0.08037 mol L^{-1}, 36.6 mN m^{-1}, 24.8 °C 0.08713 mol L^{-1}, 36.7 mN m^{-1}, 24.7 °C

the business context. Recently, there has been some controversy about the terms *data-driven* and *data-informed*. Whereas the former suggests letting data guide the decision-making process, the latter understands data as a mere plausibility check on intuition. In that sense, data should be used to substantiate your intuitions or beliefs concerning a specific subject. The other way around, data contradicting your thoughts and beliefs is to be understood as an occasion to reconsider them. If conducted properly, experimental reality hardly ever lies. In some situations, it may be wise to rethink relationships deemed certain in light of newly acquired experimental evidence. In a way, this is what *data-informed* means.

– How to respond to new questions once the data collection has been completed? Feeling the urge to look at a certain experiment from another perspective is a very common pattern—and perfectly natural. Once a certain analysis has been carried out, we learn something from the results, get insights and develop new ideas about what else could be hidden in the data. In a later chapter of the book, extraction of additional parameters and their incorporation in an existing structure will be demonstrated using the briefly introduced example of surface tension isotherms.

With these challenges and questions described, the stage is set to dive into more practical aspects. In the following, a step-by-step tutorial for data management of exemplary surface tension data will be given. For the purpose of data handling, processing, analysis and plotting, we will rely mostly on Python.

2 Python quick start

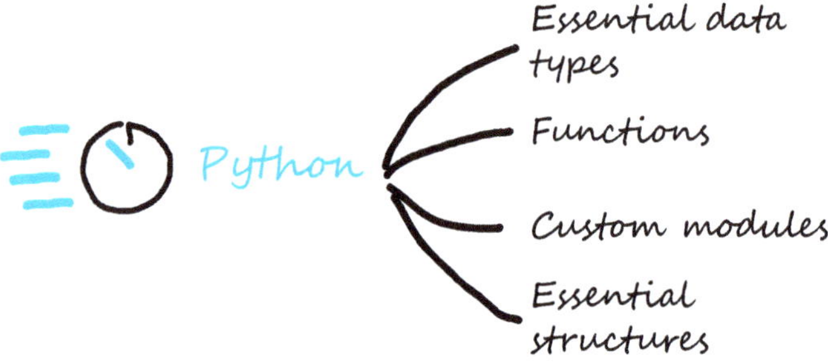

2.1 Why Python?

Throughout this book, *Python* is the tool of choice for all the tasks associated with data treatment, manipulation and basic visualization. Let me explain, why this might also be the right approach for your project.

On the Python homepage,[1] it is described as a *programming language that lets you work quickly and integrate systems more effectively*. Anyone who has read a few lines of Python will agree that the programming project's philosophy of focusing on *code readability* and readily *understandable syntax* has clearly been met—at least in comparison to other programming languages. To reflect this characteristic, the term *pythonic* has been coined. This essential feature provides a rather low entry barrier to Python: both for programming beginners and more experienced programmers aiming to engage in a new programming language. Furthermore, it is an open-source project with a large community of active users and contributors. At the time of the writing of this book, Python consistently ranks as one of the most popular programming languages.[2]

Further benefits of Python include its versatility provided by dedicated libraries for, among others, data science, scientific computing, image processing, computer vision, web programming and scraping and many more. Also, the availability of extensive documentation is an additional aspect in favor of Python.

2.2 Starting up

Before actually starting with the first coding examples, I would like to familiarize you with some options to get Python up and running on your system. Please be aware that the examples shown throughout this book are run on a Windows platform. As a first

1 https://www.python.org/

2 https://en.wikipedia.org/wiki/Python_(programming_language)

https://doi.org/10.1515/9783110788433-002

step you can download one of the many *distributions* of Python. These include the Python programming language, commonly used *packages* and an *integrated development environment (IDE)*. Personally, I have worked a lot with

- Anaconda (https://www.anaconda.com/) and
- WinPython (https://winpython.github.io/).

For installing Anaconda, navigate to the download area (https://www.anaconda.com/products/distribution), select your operating system and follow the installation wizard. In the timeframe of a few minutes, the process should be completed. In order to verify the installation of Python, go to the command line and type in

```
python --version
```

or

```
python -V
```

Either way, the response should be the version number of your Python installation, e. g., 3.9.1. The examples shown in this book were developed with the Python 3 version.

Concerning the previously mentioned *packages* coming with a distribution, roughly speaking, they serve to extend the function and capability stack of Python alone. There are, for example, several packages allowing more or less interactive plotting of data, packages for machine learning, dedicated packages for processing of natural scientific data and also computer vision. The list can be continued (almost) infinitely and is ever growing. In case you just cannot find the right module for you, there is also the option to write custom packages and make them available to the community via the Python Package Index PyPI.[3]

The IDE coming with both Anaconda and WinPython is called *spyder*. Compared to other options, spyder offers less features in exchange for a well-arranged interface. Nevertheless, it is an excellent choice for the purpose of scripting as shown herein. A screenshot of spyder is shown in Figure 2.1.

Thanks to the option of running individual lines of code or larger sections of your code contained in *cells*, a significant part of the job can be done—note—without debugging (see the following definition), as you get an immediate response.

Debugging is the process of locating and removing coding mistakes in computer programs. In information technology and engineering, the word *bug* is a synonym for the word *error*. The goal of debugging is to identify and correct an error's root cause.[4]

---

3 https://pypi.org/

4 https://www.techopedia.com/definition/16373/debugging

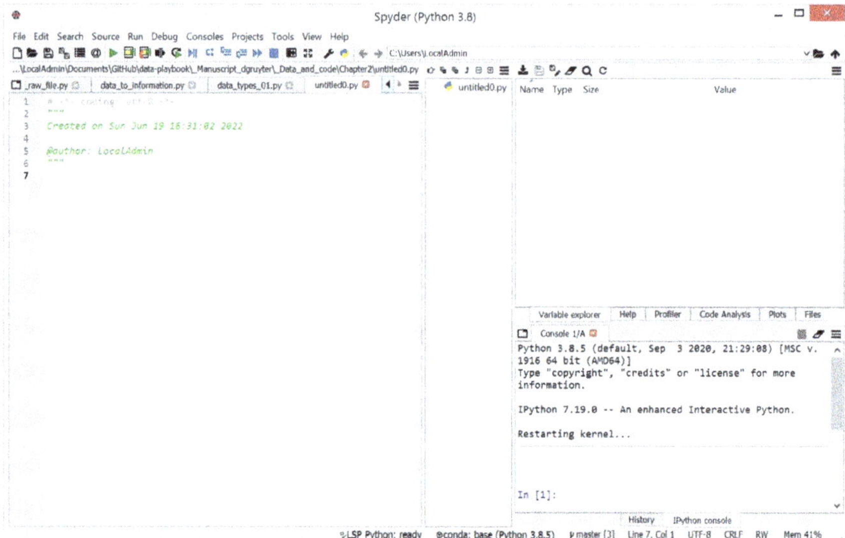

Figure 2.1: Screenshot of the IDE *spyder* coming with the Python distributions Anaconda and WinPython. The large pane on the left-hand side shows the scripting area. It represents a text editor having additional Python-specific syntax highlighting and auto-completion features. The *console* window on the bottom right allows running Python code directly. We will use this in the following to access help and information on functions and other Python objects. On the top right, several options for displaying relevant information related to a script are accessible. In particular, the *Variable explorer* will be introduced in the following section. As for most other IDEs, *spyder* also allows for customization of its appearance. Windows can be rearranged and colour schemes can be set according to user preferences. A new file (also referred to as a *script*) is generated using the blank sheet icon on the top left. The code developed in the file can be run either entirely using the play button icon (`F5`), within the limits of a cell (`Ctrl+Return`) or just in a single line (`F9`). These actions are also available via the menu entries *File* and *Run*, respectively.

---

i Individual lines of codes in spyder are run by marking the lines and pressing `F9`. A cell is opened in the code via *% ##* and run by pressing `Ctrl+Return` simultaneously. This method is frequently used for running the following code snippets.

With this basic knowledge of the IDE, we can start spyder, apply these tips and write a first tiny script to get to know the *Zen of Python* by importing the `this` package. In the following, code snippets will be shown *verbatim*, i. e., exactly as in the IDE.

```
# import (one line comment)
import this

# I am a multiline
# comment.

```

```
"""
NOT RECOMMENDED: Use of docstring
as a multiline comment.
"""
```

One-line comments in Python (shortcut `Ctrl + 1` in the IDE spyder) begin with #. As there is not built-in support in Python for multiline comments, consecutive multiline comments (shortcut `Ctrl + 4` in spyder) are recommended.[5]

Note, that in order to run this script as a cell, you do not even have to save the content written in the temporary script opened upon starting spyder. In the console window, you will get the following response when importing the `this` package:

```
The Zen of Python, by Tim Peters

Beautiful is better than ugly.
Explicit is better than implicit.
Simple is better than complex.
Complex is better than complicated.
Flat is better than nested.
Sparse is better than dense.
Readability counts.
Special cases aren't special enough to break the rules.
Although practicality beats purity.
Errors should never pass silently.
Unless explicitly silenced.
In the face of ambiguity, refuse the temptation to guess.
There should be one-- and preferably only one --obvious way to do it.
Although that way may not be obvious at first unless you're Dutch.
Now is better than never.
Although never is often better than *right* now.
If the implementation is hard to explain, it's a bad idea.
If the implementation is easy to explain, it may be a good idea.
Namespaces are one honking great idea---let's do more of those!
```

When in doubt of how to approach a certain issue, keep these memorable propositions in the back of your head. It is possible that a solution is just around the corner based on these principles.

---

5 The use of a *docstring* (indicated by triple quotes) is a possible workaround for "true" multiline comments as is seen in other programming languages. This, however, is not recommended because a *docstring*'s purpose is to provide some information on functions and not to serve as a way to comment.

2.3 Essential data types

2.3.1 Integer, float and string

Next to a basic understanding of the IDE, a sound knowledge of the most relevant data types of Python is required. Defining *integers*, *floats* and *strings* is as easy as declaring the name of the respective variable and specifying a value. Clearly, those are not the most heavily used data types within this book on this *atomistic* level. Nevertheless, they serve as *the* essential building blocks in many of the derived more complex data structures introduced in the following and can potentially lead to problems, i. e., unexpected behaviour therein if not used properly.

! Python is a *dynamically typed* programming language. Consequently, the `type` of a variable has not to be assigned to the variable before actually using it. Furthermore, the `type` of a variable is allowed to change multiple times over its lifetime. Python is also an *object-oriented* programming language. Every variable in Python is an object. In essence this means that each variable can have `attributes` and `methods` attached to it. Examples will be shown in the following parts of this book.

There is no need to explicitly declare the type of the variable. In order to verify the creation of variables of the appropriate type, we can check a variable's type via the built-in function `type`. Of course, this verification is not necessary, but, for the beginning of your Python journey or in case of working with new and therefore unknown objects, checking the type can come in handy.

Listing 2.1: data_types_01.py

```
# define integer named "my_int" with value "1"
my_int = 1

# define float named "my_float" with value "1.9912531"
my_float = 1.9912531

# define string named "my_string" with value "Hello, world"
my_string = "Hello, world"

# get type of the variables and print information
print("my_int is of type", type(my_int))
print("my_float is of type", type(my_float))
print("my_string is of type", type(my_string))
```

In spyder, there's also a window showing *Name*, *Type*, *Size* and *Value* of all variables used in the respective script. This is trivial for the variables just shown, but the *Variable explorer* shown in Figure 2.2 becomes more interesting for the data types introduced later.

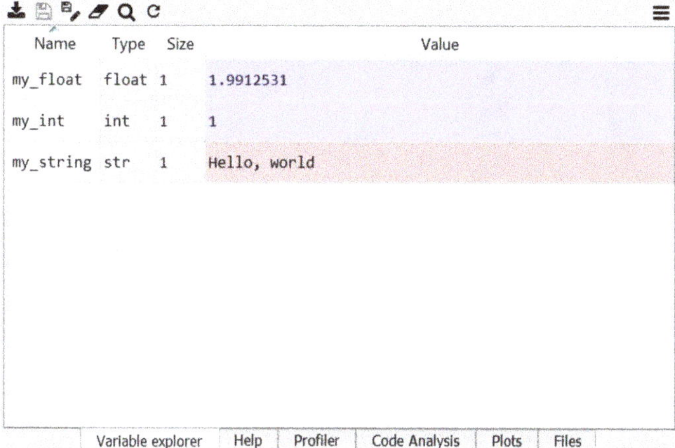

Figure 2.2: Screenshot of the *Variable explorer* in the spyder IDE. It summarizes information of *Name*, *Type*, *Size* and *Value* of all variables used in the current script.

One remark relating to strings: They can be defined both via single and double quotes.[6] In most cases, the only aspect to consider is to use the same delimiters at the beginning and end of a `string` definition.

Listing 2.2: data_types_01.py (continued)

```
# define string named "my_second_string"
my_second_string = 'I am also a valid string!'

# print string
print(my_second_string)
```

A further data type particularly relevant for natural scientists is *complex numbers*. Those are also readily available in Python. They can be specified either as the sum of a real and an imaginary part declared via `j` or as the two arguments of the `complex` statement. In addition, calculations of complex numbers is built-in as shown in Listing 2.3.

Listing 2.3: data_types_01.py (continued)

```
# defintion of complex numbers "a" and "b"
a = 2 + 2j
```

6 It is considered best practice to use double quotes for natural language messages. In cases where single quotes are part of the `string`'s content, double quotes are needed for its definition.

```
b = complex(-1, 1)

# sum of complex numbers
a_plus_b = a + b
# show results
print(a_plus_b)
```

In chemistry and material science applications, complex numbers are relevant in many fields.

The print command

Getting fast feedback on what actually happens during the execution of a script is helpful for the identification of undesired behaviour. One of the "cheapest options" is probably the introduction of `print` commands. Since the release of Python 3.6, *f-strings* serve as a new, faster and more readable way to format strings.

In short, those *formatted string literals* have an "f" at the beginning and curly braces containing expressions that will be replaced with their values. An example of using an f-string alongside the previously recommended `.format` syntax is shown in Listing 2.4.

Listing 2.4: print_strings.py

```
# define variables
sport = "Table Tennis"
ball_diameter_mm = 40

# print information using "f-strings"
print(f"The ball diameter in {sport} is {ball_diameter_mm} mm.")

# print information using "str.format"
print("The ball diameter in {} is {} mm.".format(
        sport,
        ball_diameter_mm
        )
    )
```

The console yields the following output:

```
The␣ball␣diameter␣in␣Table␣Tennis␣is␣40␣mm.
The␣ball␣diameter␣in␣Table␣Tennis␣is␣40␣mm.
```

2.3.2 List

The first more complex data type I would like to introduce is the Python `list`. As the name suggests, it is a list of objects. Python `lists` can hold different types of elements. Keep in mind that Python is an object-oriented programming language, and each variable is an object. For the subjects covered within the scope of this book, however, it is highly recommended to use `lists` containing only one type of variables, i. e., just strings, or integers or floats. There are two ways of declaring lists:
- declaring a list with all its items or
- starting with an empty list and appending items.

In order to get the number of items in a list, just use the `len`-function and specify the list as an argument.

Listing 2.5: data_types_02.py

```
# define an empty list named "shopping_list"
shopping_list = list()
# append items to the list
shopping_list.append("shoes")
shopping_list.append("hat")
shopping_list.append("3 light bulbs")
shopping_list.append("bananas")

# info on number of items on the list
print(f"There are {len(shopping_list)} items on the shopping list.")
```

The example code of Listing 2.5 already highlights an important property of a Python `list`: It is *mutable*. This means that this types of the variables are able to and intended for modification after their initial creation.

Now that we have come to lists, it is time to show the first *pythonic* feature: looping through lists. This is both easy to read and probably one of the most frequently applied techniques throughout this guide. Let's get started with printing a list of exemplary lotto numbers according to Listing 2.6. Please note that the true *pythonic* way of going through the elements of a list is the more shorthand, one-line code shown therein.

Listing 2.6: data_types_02.py (continued)

```
# define a list with all items
my_lotto_numbers_list = [13, 15, 8, 38, 2, 17]

# information setup
print("My lotto numbers are:")

```

```
# loop through the list of numbers
for number in my_lotto_numbers_list:
    # print the number
    print("  -", number)

# remove "looping variable" named "number"
del number

# loop through the list of numbers ("pythonic")
print("My lotto numbers are:")
[print(f"  - {number}") for number in my_lotto_numbers_list]
```

Either way, the console yields the following output:

```
My␣lotto␣numbers␣are:
␣␣-␣13
␣␣-␣15
␣␣-␣8
␣␣-␣38
␣␣-␣2
␣␣-␣17
```

Furthermore, accessing the individual elements of a `list` is achieved via *indexing*. For this purpose, place the number of the element you wish to access in square brackets.

Within the definition of the loop, the variable `number` is defined implicitly and can be used for further tasks. This can be particularly helpful in the course of developing scripts for analysing experimental data. In case you do not want to see the *looping variable* once the loop terminated, there are two options:

- remove the variable using the `del` statement or
- make the looping variable invisible right from the beginning by starting its name with an underscore.

---

i Variables names beginning with a literal underscore, "_", will *not* be shown in the spyder IDE's *Variable explorer* by default. This will also lend clarity to the *Variable explorer*, especially when working with a large number of variables.

The elements of a list are accessed via indexing with square brackets. The index starts at zero. In order to address the first element of a list, you would use index zero for the second element, use index one, etc. An example of accessing explicitly the first element of the `shopping_list` is given in Listing 2.7.

Listing 2.7: data_types_02.py (continued)

```
# access the first item on the "shopping list"
first_item = shopping_list[0]
# info
print("The first item on the list is:", first_item)
```

A somewhat more advanced option for looping through a `list` is to use the `enumerate`-function. This will return two variables: the index (variable `idx`) and the value (variable `number`) of the accessed element. As the indexing starts at zero, the value 1 is added in order to show lottery ball numbers from 1 to 6 in Listing 2.8.

Listing 2.8: data_types_02.py (continued)

```
# loop through the list of numbers via "enumerate"
for idx, number in enumerate(my_lotto_numbers_list):
    # print the index and number
    print(f"Ball #{idx+1}: {number}")
```

2.3.3 Tuple

A data type very similar to a Python `list` is the `tuple`. In the same sense they are also ordered and can be indexed in a similar manner as lists. But, in contrast to lists, they are *unchangeable* or *immutable*. This means that a `tuple` cannot be changed, and items cannot be added or removed after the `tuple` has been created as easily as for `lists`. According to the documentation, a `tuple` is a built-in immutable sequence. Nevertheless, there are some handling operations left if you really need to modify the contents of a `tuple` as indicated in Listing 2.9. Therein, we define the tuple `my_tuple` holding the strings `a`, `b` and `c`. Next, the second entry of `my_tuple` is printed before adding the characters `d` to `g` and `x` to the tuple. Indexing for `tuples` is demonstrated by printing the last four entries. Finally, the element `x` is replaced by `h` by transforming the `tuple` to the previously introduced `list` and back again to a `tuple`.

The documentation can be accessed via the spyder console (bottom right area in Figure 2.1) by typing in the function name, followed by ?, i. e., `list?`.

Listing 2.9: data_types_05.py

```
# define a tuple
my_tuple = ("a", "b", "c")

# get second item via indexing
```

```
print("Second item is", my_tuple[1])

# add further letters to the tuple
my_tuple += ("d", "e")

# add one more letter to the tuple one by one
my_tuple += ("f",)
my_tuple += tuple("g")

# add another letter (long version of "+="-notation)
my_tuple = my_tuple + tuple("x")

# print last 4 items in the tuple / "slicing" to show the introdcued
# change (modified tuple)
print(my_tuple[-4:])

# change type to list in order to make changes
my_list = list(my_tuple)

# make the change/correction: letter #8 is "h"
my_list[7] = "h"

# get back to immutable tuple --> get tuple from list and overwrite
# existing tuple
my_tuple = tuple(my_list)
```

The console yields the following output:

```
Second␣item␣is␣b
('e',␣'f',␣'g',␣'x')
```

Go through the code of Listing 2.9 line by line to follow the changes in `my_tuple`. Closely keep an eye on spyder's *Variable explorer*.

In Listing 2.9, the `+=` shorthand assignment was introduced. Instead of modifying a variable and assigning the result to the previous variable name (`a = a + 5`), the modification and reassignment is done in one step (`a += 5`). The actions are fully equivalent. There are analogous assignments for subtraction, multiplication and division next to the shown addition type.

2.3.4 Dictionary

Python *dictionaries* are a further useful data type. In contrast to the previously introduced `lists`, they are made up of *key-value pairs*. This implies some differences when accessing a dict's content and looping through a dict. Also for dictionaries, you can start from either an empty dict, which is subsequently filled with items, or initialize it with key-value pairs right from the beginning. Just as for lists, a dict can hold values of various different types. For dictionaries, however, this is less critical because the *key* should ideally provide some clue to the *value* type to be expected. The basic functions related to dicts are summarized in Listing 2.10. Therein, we define the dictionary my_dict having the keys name, training and favorite number. After this definition step, keys and values are accessed one after the other. Also, simultaneous looping over keys and values of a dict is demonstrated. Rather than in the case of the previously shown lists and tuples, this is not done via indexing, but using the key corresponding to a value.

Listing 2.10: data_types_03.py

```
# define an empty dictionary
my_dict = dict()

# introduce information to dict
my_dict["name"] = "Matthias"
my_dict["training"] = "chemistry"
my_dict["favorite number"] = 8

# loop through only "keys"
for _key in my_dict.keys():
    # info
    print("  -", _key)

# loop through only "values"
for _val in my_dict.values():
    # info
    print(_val)

# loop through both "keys" and "values"
for _key, _val in my_dict.items():
    # info
    print(_key, ":", _val)

# access the "name" specified in the dictionary "my_dict"
print("The name is", my_dict["name"])

# remove a key-value pair
del my_dict["favorite number"]
```

The console yields the following output:

```
-␣name
-␣training
-␣favorite␣number
Matthias
chemistry
8
name␣:␣Matthias
training␣:␣chemistry
favorite␣number␣:␣8
The␣name␣is␣Matthias
```

The `key-value` type accessing of `dict`-elements makes it feel more defined compared to `lists` where you merely have the option of indexing. It is nevertheless important to have both options at hand and use them appropriately depending on the situation.

2.3.5 Data frame

To conclude the introduction of the most important types, we'll move on to the *DataFrame* contained in the *pandas* package. In short, `pandas` is probably *the* primary data analysis and manipulation tool that is built on top of the Python programming language.[7] Therefore, the `pandas` documentation should become one of your dearest friends. In particular, `DataFrames` can be used for pretty much anything—from handling to cleaning and basic visualization of data. There are many options for getting hands on a `pandas.DataFrame`. For readability, I will refer to this variable type merely as `DataFrame` in the following. As shown in the next chapters, we create a `DataFrame` via reading from different sources such as *txt*-files, *xlsx*-files or even databases. Still, you can instantiate a `DataFrame` from scratch with the previously introduced `lists` or `dicts`. To use `DataFrames` in our scripts, we have to import the `pandas` package. Almost on any occasion, `pandas` is *aliased* as `pd`. Also, you will find this shorthand notation of the package online in almost any forum or thread dedicated to `pandas`. In Listing 2.11, the variable `my_dataframe` is initialized empty before adding the columns `x` and `y` to it. After its creation, basic properties are requested and printed.

Listing 2.11: data_types_04.py

```
# import the "pandas" module as alias "pd"
import pandas as pd

# create the empty pd.DataFrame "my_dataframe"
```

7 https://pandas.pydata.org/

```
my_dataframe = pd.DataFrame()

# add columns "x" and "y"
my_dataframe["x"] = [10, 20, 30, 40, 50]
my_dataframe["y"] = [0.02, 0.10, 0.28, 0.31, 0.34]

# # alternatively: creation of DataFrame from dictionary
# my_dataframe = pd.DataFrame({
#           "x" : [10, 20, 30, 40, 50],
#           "y" : [0.02, 0.10, 0.28, 0.31, 0.34]
#         })

# get column names
print("my_dataframe consists of columns", my_dataframe.columns)

# get information on row count
print(f"my_dataframe consists of {len(my_dataframe)} rows.")
```

The console yields the following output:

```
my_dataframe consists of columns Index(['x', 'y'], dtype='object')
my_dataframe consists of 5 rows.
```

Upon initializing a `DataFrame` via `lists`, all columns are required to have the same length, i. e., contain the same number of elements in the constituent `lists`. The same holds for creating a `DataFrame` from a `dict` (shown as "alternative" in Listing 2.11.).

One immediately apparent strength of a `DataFrame` is the option to easily introduce columns derived from already existing ones. Typical use cases for this feature in the context of natural sciences are the conversion of units or offsetting the values stored in a particular column. Also, focussing on—for whatever reason——specific parts of a `DataFrame` is possible using the `query`-method of the `DataFrame`. After this very basic treatment of data, you might wish to visualize the results to get a first impression of the data. The `DataFrame` is here to help also with that via its `plot` method as shown in Listing 2.12.

Listing 2.12: data_types_04.py (continued)

```
# introduce a modified column
my_dataframe["x_plus_offset"] = my_dataframe["x"] + 5

# consider only point where "x_plus_offset" is below "44"
my_selection = my_dataframe.query("x_plus_offset < 44")

# make a scatterplot of "x_plus_offset" against "y"
```

```
my_selection.plot(
    x="x_plus_offset",  # column name of x-variable
    y="y",  # column name of y-variable
    kind="scatter"
    )

# import plotting module
import matplotlib.pyplot as plt

# labelling of axes
plt.xlabel("x plus offset")
plt.ylabel("Value of y")

# make a barplot of "x_plus_offset" against "y"
my_selection.plot(
    x="x_plus_offset",
    y="y",
    kind="bar"
    )
```

The resulting plots are shown in Figure 2.3 and Figure 2.4, respectively.

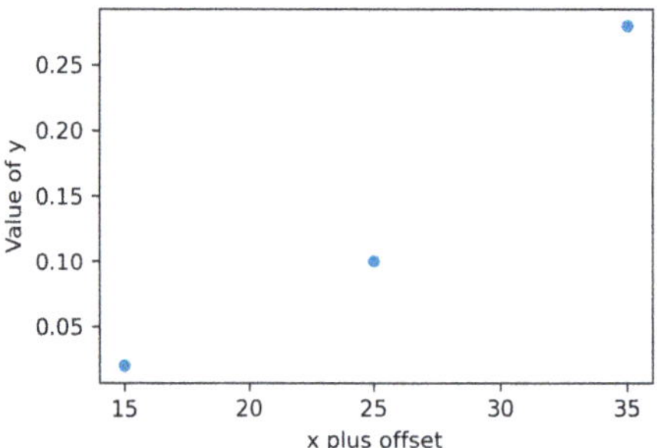

Figure 2.3: Scatter plot of exemplary data obtained via the `plot`-method of a `DataFrame`.

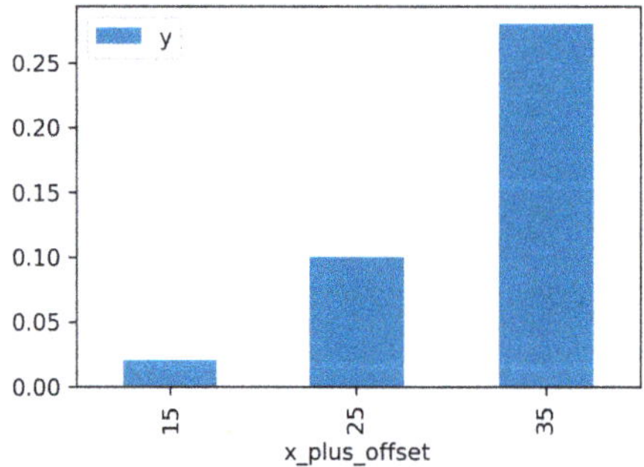

Figure 2.4: Bar plot of exemplary data obtained via the `plot`-method of a `DataFrame`.

2.3.5.1 melt and pivot

The DataFrame methods melt and pivot are counterparts for gathering columns into rows (melt) and spreading rows into columns (pivot). The full details and options of the current version are best understood via the pandas.DataFrame online documentation. A schema of the associated impact on a DataFrame is given in Figure 2.5.

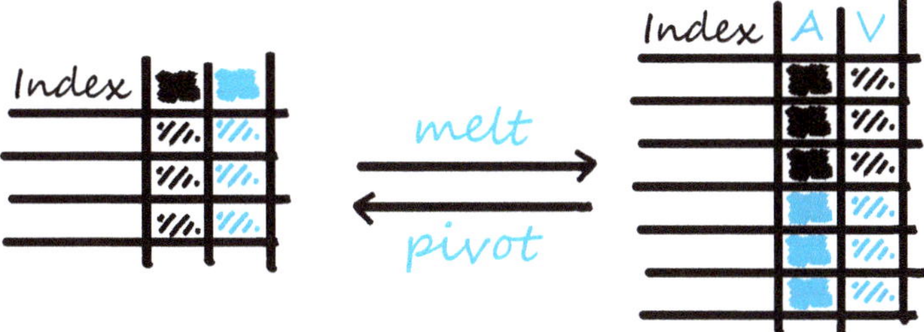

Figure 2.5: Schematic representation of the melt- and pivot-methods of a DataFrame for reshaping its content.

The basic usage of the pivot- and melt-methods alongside the restoration of the original DataFrame upon running them subsequently is shown in Listing 2.13.

Listing 2.13: melt_pivot.py

```
# import
import pandas as pd

# build exemplary "original" DataFrame
data_original = pd.DataFrame({
            "sample"              : ["A", "B", "C"],
            "mass_at_5min_g"      : [110, 108, 111],
            "mass_at_15min_g"     : [104, 105, 92]
            })

# print
print(data_original)

# "melt" original DataFrame (== "columns to rows")
data_melted = data_original.melt(
            id_vars="sample",
            value_vars=["mass_at_5min_g", "mass_at_15min_g"]
            )
# print
print(data_melted)

# "pivot" melted DataFrame (== "rows to columns")
```

```
data_melted_and_pivoted = data_melted.pivot(
            index="sample",
            columns="variable",
            values="value"
            )
# add index information as "sample" column
data_melted_and_pivoted["sample"] = data_melted_and_pivoted.index
# info
print(data_melted_and_pivoted)
```

The console yields the following output:

```
␣␣sample␣␣mass_at_5min_g␣␣mass_at_15min_g
0␣␣␣␣␣␣A␣␣␣␣␣␣␣␣␣␣␣␣␣110␣␣␣␣␣␣␣␣␣␣␣␣␣␣104
1␣␣␣␣␣␣B␣␣␣␣␣␣␣␣␣␣␣␣␣108␣␣␣␣␣␣␣␣␣␣␣␣␣␣105
2␣␣␣␣␣␣C␣␣␣␣␣␣␣␣␣␣␣␣␣111␣␣␣␣␣␣␣␣␣␣␣␣␣␣␣92
␣␣sample␣␣␣␣␣␣␣␣␣variable␣␣value
0␣␣␣␣␣␣A␣␣␣mass_at_5min_g␣␣␣␣110
1␣␣␣␣␣␣B␣␣␣mass_at_5min_g␣␣␣␣108
2␣␣␣␣␣␣C␣␣␣mass_at_5min_g␣␣␣␣111
3␣␣␣␣␣␣A␣␣mass_at_15min_g␣␣␣␣104
4␣␣␣␣␣␣B␣␣mass_at_15min_g␣␣␣␣105
5␣␣␣␣␣␣C␣␣mass_at_15min_g␣␣␣␣␣92
variable␣␣mass_at_15min_g␣␣mass_at_5min_g␣sample
sample
A␣␣␣␣␣␣␣␣␣␣␣␣␣␣␣␣␣␣␣␣␣104␣␣␣␣␣␣␣␣␣␣␣␣␣110␣␣␣␣␣␣A
B␣␣␣␣␣␣␣␣␣␣␣␣␣␣␣␣␣␣␣␣␣105␣␣␣␣␣␣␣␣␣␣␣␣␣108␣␣␣␣␣␣B
C␣␣␣␣␣␣␣␣␣␣␣␣␣␣␣␣␣␣␣␣␣␣92␣␣␣␣␣␣␣␣␣␣␣␣␣111␣␣␣␣␣␣C
```

2.3.5.2 merge

The merge-function is available both via the pandas package or as a method of an existing DataFrame. In essence, two DataFrames are combined in a defined way according to the entries in certain columns. A schematic representation is given in Figure 2.6. A full example related to our use case data is discussed in more detail in Subsection 7.2.2. Also, for the merge-function and method, the full details and options applying to your installed pandas-version are accessible via the respective online documentation.

Figure 2.6: Schematic representation of merging two DataFrames based on the values in columns A and B. There are multiple options for specifying the rows to appear in the resulting DataFrame. The herein sketched option "inner" leads to a row in the resulting DataFrame only if the value appears both in columns *A* and *B* of the original DataFrames on which the merge is carried out.

The basic usage of the merge-function is shown in Listing 2.14.

Listing 2.14: merge.py

```
# import
import pandas as pd

# build individual DataFrames to be merged
data_A = pd.DataFrame({
            "A"             : ["A", "B", "C"],
            "c_A"           : [1, 2, 3]
            })

data_B = pd.DataFrame({
            "B"            : ["A", "E", "B"],
            "c_B"          : [2, 8, 4]
            })

# print original DataFrames
print("data_A")
print(data_A)
print("data_B")
print(data_B)

# merge DataFrames on columns "A" and "B"
data_merged = pd.merge(
        data_A,  # "left" DataFrame
        data_B,  # "right" DataFrame
        left_on="A",  # column on which left DataFrame is merged
        right_on="B",  # column on which right DataFrame is merged,
        how="inner"
        )
# print merged DataFrame
print("data_merged")
print(data_merged)
```

The console yields the following output:

```
data_A
␣␣␣A␣␣c_A
0␣␣A␣␣␣␣1
1␣␣B␣␣␣␣2
2␣␣C␣␣␣␣3
data_B
␣␣␣B␣␣c_B
0␣␣A␣␣␣␣2
1␣␣E␣␣␣␣8
2␣␣B␣␣␣␣4
data_merged
␣␣␣A␣␣c_A␣␣B␣␣c_B
0␣␣A␣␣␣␣1␣␣A␣␣␣␣2
1␣␣B␣␣␣␣2␣␣B␣␣␣␣4
```

2.3.5.3 groupby

The `groupby`-method is a powerful tool, among others, for getting *summary statistics* of a `DataFrame`. Exemplarily, the calculation of a mean height by gender is shown in Listing 2.15. Alternatively, the `groupby`-method returns an iterator that allows further processing of each subset of data.

Listing 2.15: groupby.py

```python
# imports
import pandas as pd

# generate example DataFrame
names = ["Lisa", "Sara", "Michael", "Josef"]
gender = ["f", "f", "m", "m"]
height = [164, 172, 182, 177]
# lists to DataFrame via implicitly defined python dictionary {}
data = pd.DataFrame({
        "name"      : names,
        "gender"    : gender,
        "height_cm" : height,
        })

# get mean height by gender using "groupby"
result = data.groupby(by="gender").mean()
# show results
print(result)

# work with data subsets via iterator
```

```
for _gender, _data in data.groupby(by="gender"):
    # info
    print("\nData subset:", _gender)
    print(_data)
```

The console yields the following output:

```
            height_cm
gender
f           168.0
m           179.5
Data subset: f
   name gender height_cm
0  Lisa      f       164
1  Sara      f       172
Data subset: m
      name gender height_cm
2  Michael      m       182
3    Josef      m       177
```

2.4 Essential structures

In Python the number of spaces at the beginning of a code line (indentation) is critical. Whereas indentation is relevant for readability only in most other programming languages, Python uses it to indicate blocks of code. Indentation is relevant for many of the concepts introduced in the following. In cases where difficulties are encountered reproducing the code following snippets and getting them to run, indentation might be a possible source of error.

Almost any IDE used for the development of Python scripts and programs will provide some assistance in conforming to the indentation requirements. Additionally, Python provides instructive error messages.

2.4.1 `if`-conditions

Reacting to specific conditions is necessary to direct a script in particular directions depending on, e. g., input variables or other “critical values”. For this situation, Python provides case selection via the keywords `if`, `elif` and `else`. An example is given in Listing 2.16. Therein, we define the `str` variable `my_name` and check if the value equals a certain value. Depending on the results of this check, a different response is obtained.

Listing 2.16: if_elif_else.py

```
# define string variable
my_name = "Matthias"

# my_name = Christian? (check for eqaulity via == operator)
if my_name == "Christian":
    # message
    print("Hello, Christian.")
# my_name = Matthias? ("elif" reads as "else if")
elif my_name == "Matthias":
    # message
    print("Hello, Matthias.")
# any other case
else:
    # message
    print("Hello, dear user.")
```

The console yields the following output:

```
Hello,␣Matthias.
```

! To check for equality, the ==-operator is applied.

2.4.2 for-loops

As already introduced in Listing 2.1, *going over lists* or any other type of `iterable` is achieved using `for`-loops. In addition to the `for` statement, there are at least two other commands useful within. `continue` makes the for-loop go to the next element immediately, while `break` terminates the iteration over the list. A schematic representation is given in Figure 2.7, an example in Listing 2.17. Therein, a `list` of tools is defined

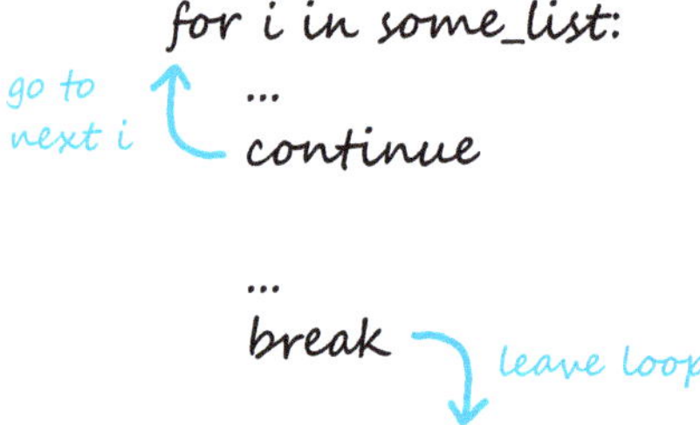

Figure 2.7: `continue` and `break` in `for`-loops. The former directs the code to the next element, the latter terminates the loop. In order to obtain a better understanding of the individual steps during the execution of loop, the visualization provided by https://pythontutor.com/ is recommended.

in the variable `my_list`. We loop over the list and print the elements in it with some additional constraints. We immediately move to the next element if the present element is `nail` and terminate the looping if the present element is `saw`. In both of the latter cases, the current tool will not be printed since the respective function is called only at the end of the loop and accordingly *not* reached in the case of any preceding `continue` or `break`.

Listing 2.17: for_continue_break.py

```
# list of examples
my_list = ["hammer", "nail", "saw", "screw driver"]

# go over list
for tool in my_list:

    # go to next tool, if tool is "nail"
    if tool == "nail":
        # go to next tool in the list
        continue

    # end for loop, if tool is "saw"
    if tool == "saw":
        break

    # print name of tool in any other case
    print(tool)
```

The console yields the following output:

```
hammer
```

2.4.3 zip

If you are facing the task to simultaneously loop over two or more lists, use the `zip`-function. In Listing 2.18, the joint iteration of a list of users with their corresponding default directories for saving data is shown as an example. Explicitly, experiments corresponding to user `Alexander` are to be stored in a folder `productive`, the data of user `Matthias` in a folder named `test`.

Listing 2.18: zip.py

```
# specify lists of users and directories
users = ["Alexander", "Matthias"]
directories = ["productive", "test"]
```

```
# loop list simultaneously
for u, d in zip(users, directories):
    # show information
    print(f"Data of user {u} are saved to directory {d}.")
```

The console yields the following output:

```
Data␣of␣user␣Alexander␣are␣saved␣to␣directory␣productive.
Data␣of␣user␣Matthias␣are␣saved␣to␣directory␣test.
```

2.4.4 Functions

Functions are an integral part in our toolbox for doing all possible kinds of things related to experimental data—from reading files to processing data in one way or the other, or storing processed results. If there is no apparent out-of-the box solution, a custom defined `function` might be just the right thing for your purpose. A `function` is a part of code that runs when *called*. In order to spice things up, a function can take multiple arguments to be processed and one or more results can be returned. Typical `return` values within the scope and use cases of this book include processed `DataFrames` or *flags*.

> A *flag* is a value that acts as a signal for a function or process. A flag's value helps, e. g., in determining the state in which a function returns.

A basic function without any input arguments for just printing a text is shown in Listing 2.19. A function definition is initialized with the `def` statement and terminated with `return` alone, or the one or multiples names of the Python objects to be returned. If there is nothing after the `return` statement, the function will return the `None` value.

> A `return` at the end of a function definition is not required unless you want to actually *use* a value "returned from" the function. If not explicitly specified, a function will return `None`.

In this basic example, the function `say_hello` is called once individually and five times via a `for`-loop.

Listing 2.19: functions_01.py

```
# define function "say_hello"
def say_hello():
    # carry of the functions task
```

```
    print("Hello (from the frunction).")
    # return
    return

# call the function
say_hello()

# say hello 5 times
print("Hello from loop:")
for i in range(5):
    # call function
    say_hello()
```

The console yields the following output:

```
Hello␣(from␣the␣frunction).
Hello␣from␣loop:
Hello␣(from␣the␣frunction).
Hello␣(from␣the␣frunction).
Hello␣(from␣the␣frunction).
Hello␣(from␣the␣frunction).
Hello␣(from␣the␣frunction).
```

According to the convention, Python functions should be named using all lowercase with underscores to increase readability.[8] The same convention holds for variable names. Furthermore, making use of a `docstring` is both highly recommended and furthermore practical if you use a custom defined function some days, weeks or even months after touching it the previous time. For example, let's define a function, which doubles an input value, if the input value is of type `float` or `int` and return nothing otherwise.

Listing 2.20: functions_01.py (continued)

```
# define properly formatted and documented function
def double_value(value, show_info=True):
    """
    returns the double value of parameter "value" if possible,
    else None.

    Parameters
    ----------
    value : int, float
```

8 https://www.python.org/dev/peps/pep-0008/

```
        value to be doubled.
    show_info : bool, optional
        flag for showing additional information during function run.
        The default is True.

    Returns
    -------
    doubled_value :
        doubled value of input "value"
    doubled_value_type :
        type of "doubled_value"

    If not appropriate input type (other than int, float),
    return None
    """

    # check type of "value" input
    if type(value) not in [float, int]:
        # Info-Message
        print("Please, use input arguments of type int or float.")
        # no doubling possible --> "leave" function here / exit from
        # this function with return value "None"
        return

    # "else"-behavior, if doubling is possible
    doubled_value = 2*value
    doubled_value_type = type(doubled_value)

    # info
    if show_info:
        print("result:", doubled_value)

    # return doubled value and its type
    return doubled_value, doubled_value_type


# run "double_value" function with different arguments
double_value(4)
double_value("4 bananas")
# call "double_value" function and store return in variable "result"
result = double_value(2.71, show_info=False)

# get "value" from variable named "results" of type "tuple"
print("Value:", result[0])
# get "type" from variable named "results" of type "tuple"
print("Type: ", result[1])
```

The console yields the following output:

```
result: 8
Please, use input arguments of type int or float.
Value: 5.42
Type:  <class 'float'>
```

A very interesting feature of Python functions is their ability to `return` multiple values. In our previous example, shown in Listing 2.20, those are the doubled value of the input and its type, if applicable. They are returned together as a `tuple` and can be accessed individually via the previously described indexing via square brackets. As already mentioned, the immutable nature of the returned `result` variable (the type of `result` can be verified as a `tuple` from spyder's variable explorer) is a helpful feature for the return value of a function: Unintentional modification is almost impossible. Of course, you can and should restrict your functions to do *exactly* one thing and also make them return just one thing to keep things clean.

2.4.5 try and except

As is to be expected from a construct named *try* and *except*, they enable checking a block of code for errors. If the `try`-block fails, your script or program will move on the `except`-block. You can think of `try` and `except` as the "more tolerant brother or sister" of `if` and `else`. If the `try`-block fails, you will run into an `Exception` about which you can get information.

For example, a `NameError` is raised when we try to use an undefined variable. In contrast to an `if`-clause, our script will not fail but rather go to the alternative pathway, i. e., the `except`-block. A basic example of this behaviour is shown in Listing 2.21, where we try to append a `DataFrame` named `data_to_append` to a—within the `try`-block—a nonexistent `DataFrame` named `data`.

Listing 2.21: try_except.py

```python
# import DataFrame
from pandas import DataFrame

# define DataFrame to be appended
data_to_append = DataFrame({
    "name" :  ["Walter", "Jesse"],
    "location" : ["Munich", "Berlin"]
    })

# try: append "data_to_append" to so far non-existing "data"
try:
    data = data.append(data_to_append)

```

```
# if not successful: define empty "data" and append to this
except Exception as e:
    # print information on Exception
    print(type(e).__name__, e)

    # define empty DataFrame (to be appended to)
    data = DataFrame()
    # append "data_to_append" to "data" (initialized empty)
    data = data.append(data_to_append)

# info
print(data)
```

The console yields the following output:

```
NameError name 'data' is not defined
     name location
0  Walter   Munich
1   Jesse   Berlin
```

We will use this pattern in some of the later scripts, in particular for appending to `DataFrames` within loops. A use case is combining equally structured data collected across multiple experimental raw files into an overall `DataFrame`.

2.4.6 User defined packages and modules

Whenever you feel that a Python function is no longer an appropriately sized *box* to collect your envisioned functionality, a custom Python `module` might be the next larger container to collect your functions in. For example, a module could be used to collect multiple functions related to raw data processing and handling of files as obtained from different devices measuring the same experimental quantity.

Furthermore, the level of a Python `package` is recommended. The `package` level is a way of structuring the namespaces coming with Python `modules`.

In order to show the mechanics of creating a `package` holding several `modules`, we will set the scientific questions aside for the moment and define a `package` named *holiday_greetings.py*. Therefore, we create a folder named *holiday_greetings* in the very directory in which we want our Python script to run. Additionally, we have to create an empty Python-file called *__init__.py* in this folder. The file *__init__.py* is required to make Python aware of the `package` and `module` structure we intend to build. Within the `package`, we can define multiple `modules`, i. e., some kind of code library containing functions, classes or basically anything you can think of to be used in your projects. In our example, let's further assume that we want to define seasonal greeting functions from different countries. This means that we define a `module` for each country. A conceptual sketch of the suggested folder structure is given in Figure 2.8.

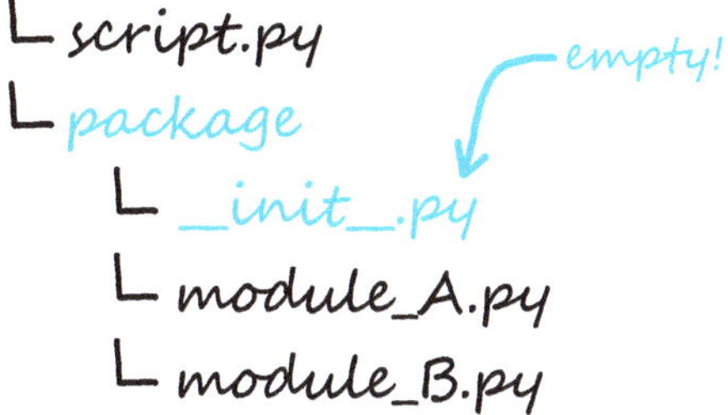

Figure 2.8: Conceptual sketch of the folder structure for defining custom `modules` *module_A.py* and *module_B.py* contained in the custom `package` to be used in *script.py*. The empty file *__init__.py* makes Python identify the `package`/`module` structure.

In order to have the messages separated by countries, we will further create files named *from_spain.py* and *from_france.py*. Those are the `modules` of our basic example. Within each of these files, we will define a function `wish_merry_christmas` printing a nice message in the respective language. A representative country specific example is given in Listing 2.22.

Listing 2.22: from_spain.py (for example `module_A.py` from Figure 2.8)

```
def wish_merry_christmas(name):
    """
    wishes a merry christmas to "name"

    Returns
    -------
    None.

    """

    # print merry christmas message
    print("Feliz Navidad,", name)
```

Now to the interesting part. In our Python script, *script.py*, we can use all of the functions defined within our `package`'s `modules`, leaving us with a much cleaner, more separated and better maintainable piece of code. To get the best wishes from Spain and France, we just have to import the relevant `modules` from our newly defined `package` named `holiday_greetings` and call the functions as shown in Listing 2.23.

Listing 2.23: modules_01.py

```
# import seasonsal greetings function from each (available) country
from holiday_greetings import from_spain, from_france

# greeting to Tobias from Spain
from_spain.wish_merry_christmas("Tobias")
# ... and to Rob from France
from_france.wish_merry_christmas("Rob")
```

The console yields the following output:

```
Feliz␣Navidad,␣Tobias
Joyeux␣No\"{e}l,␣Rob
```

Although this might seem and probably is somewhat over-engineered for the task of wishing happy holidays in this example, having some additional levels for structuring parts of your scripts will be helpful as they grow and further functionalities and options are added bit by bit. In order to counteract this *organic growth* problem, packing things together right from the beginning will be appreciated by your future self at a later stage in the journey.

The presented separation of tasks via `functions`, `modules` and `packages` is particularly helpful for the purpose of testing individual parts of your processing pipeline in response to possible sudden changes in file formats or the like. Tracing down the errors will be much easier if the overall task is split into several reasonably sized functions. Whatever that means depends on your perception, experience and aptitude to extensive debugging. Furthermore, maintaining and adjusting your code will thus be more convenient.

Using a *version control system* is strongly recommended if you wish to dive deeper into developing analysis scripts or any other kind of code (see chapter C).

! Breaking your code to small units having a defined purpose is the basis for *unit testing*. Therein, individual units or functions of software are tested independently. A unit is the smallest testable part of an application. It mainly has one or a few inputs and produces a single output.

Another benefit of organizing your functions within `packages` and `modules` is the possibility to readily share your work with colleagues from the lab—or even the entire Python community. Packaging is carried out via another package, `setuptools`. After successful completion, you will end up with a *.tar.gz*-file. This type of file can be used to install our custom `package` using the package manager `pip`. After this step, you will be able to use your newly created `module` in any script without having to copy the folder structure shown in Figure 2.8. Using the familiar `import` syntax is sufficient after installation of the custom `package`.

Exemplarily, the conversion of the herein developed code to a shareable package `surface_tension` is shown in Appendix A.

This approach also contributes to leveraging the full power of *RStudio* as "A Single Home for R & Python".[9] *R* is a programming language and environment for statistical computing and graphics that is also very popular among scientists. Accordingly, many data science teams today are "bilingual", using both *R* and Python in their work since both languages have unique strengths. In short, once developed in Python, `packages` and `modules` can be made available to users of *R* using *RStudio*.

2.5 Wrap up

This chapter introduced Python as an easy-to-learn programming language for both beginners and experienced programmers. It covered a brief rationale of why Python is used for the purposes covered in this book. Also, installation pathways of the various distributions were recommended. Additionally, an overview of the most important data types for the scope of this book including

- `integers` (Python built-in),
- `floats` (Python built-in),
- `strings` (Python built-in),
- `lists` (Python built-in),
- `tuples` (Python built-in),
- `dicts` (Python built-in) and
- `DataFrames` (from the pandas module)

was given. Furthermore, standard operations, such as looping over `lists` and `dicts`, exception handling via `try` and `except` constructs and `if-elif-else` conditions were introduced. The chapter concludes with the definition of custom `functions` and the organization of code with the help of `modules` and `packages`.

9 https://www.rstudio.com/solutions/r-and-python/

3 The steps of data processing

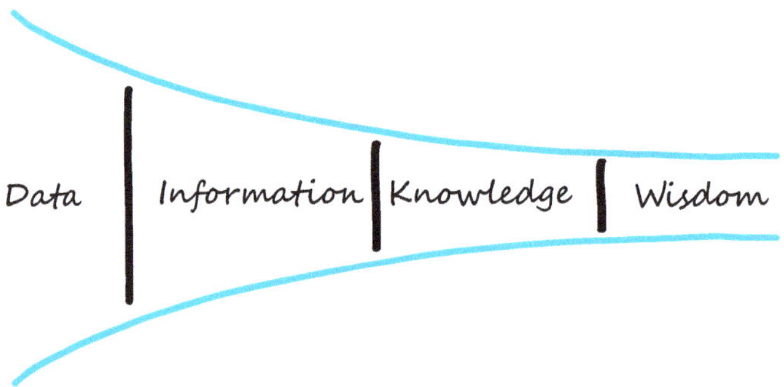

In very abstract terms, the objective and challenge of any scientific endeavour and mostly any other field is moving up the *data pyramid*.[1] This is also known under the acronym DIKW, Data—Information—Knowledge—Wisdom.

There are numerous interpretations out there omitting one or another level or introducing further intermediate stages. To better elaborate on the concept, I would like to cite one of the ways to read the model:

Information is defined in terms of data, knowledge in terms of information, and wisdom in terms of knowledge.

The strongly resonant representation as a pyramid points out the key understanding that there is no rational way to attain *knowledge* or èven *wisdom*, if the lower hierarchy levels *information* and *data* are discarded. Another way to understanding the data pyramid shown in Figure 3.1 is by identifying the things required to move from one level of the hierarchy to the next:

– data given context becomes information,
– information given meaning becomes knowledge, and
– knowledge given insight becomes wisdom.

It is evident that moving up the pyramid correlates with increasing value—be it for a research institution or a company. Whereas the bottom levels *data* and *information* make it possible to allow answer the question *what*, *knowledge* and *wisdom* pursue the *how* and *why*, respectively. The scope of this is book is, by and large, limited to the lower stages of the data pyramid.

The *wisdom* level differs from the other parts of the pyramid in a critical aspect. Rather than being directed to past findings and understandings, it involves addressing questions related to the future that are based on pre-existing knowledge and a

1 https://en.wikipedia.org/wiki/DIKW_pyramid

https://doi.org/10.1515/9783110788433-003

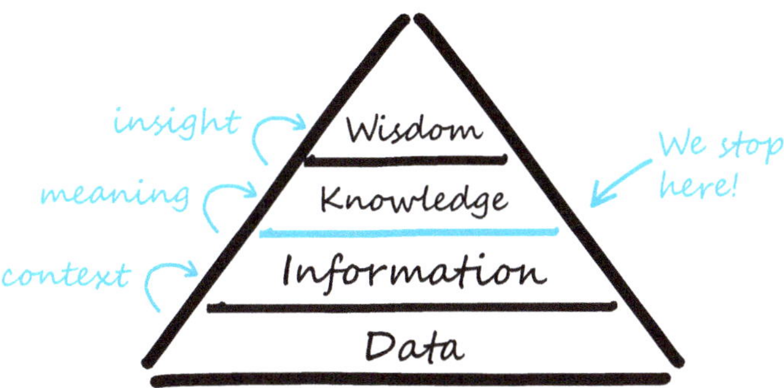

Figure 3.1: The DIKW-pyramid. No wisdom without knowledge, no knowledge without information, no information without data. The scope of this book's discussion deals primarily with the *data* and *information* levels. In Chapter 9, we provide an outlook to approaching the upper levels.

sound understanding of the latter. In short, wisdom ensures *actionable* decisions for the future based on knowledge from past experiences. Formalizing the understanding of causality, i. e., the way of reaching the topmost levels of the data pyramid, is expected to significantly contribute to attaining *wisdom* in many areas [3].

3.1 Introductory data pyramid example

Before attempting to shed light on some typical misconceptions about data analysis, I would like to explain the DIKW-model using a readily understandable example outlined in Figure 3.2.

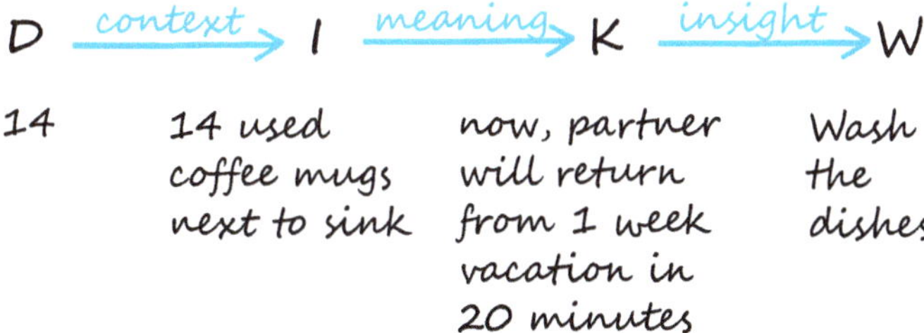

Figure 3.2: A real-life example of the DIKW-model.

Given the number 14 (data) alone, there is no emotion associated with it. Adding some context to this pure number, i. e., 14 is the number of used coffee mugs piling up next to the sink, leaves you with the information that there is a significant number of dishes in a certain state in a certain location. Let's add some meaning to this information. You are aware that your partner will return home from a one-week trip, and the situation in the kitchen turns out to be as just described. With this knowledge (14 used coffee mugs next to the sink, and the prospect of your partner returning home from a long trip hoping to find the house in a more or less acceptable condition) available, and the insight you have gained from similar situations in the past, you arrive at wisdom: an actionable recommendation for the—here immediate—future. You want to do a favour to your partner and yourself: You wash the dishes—or at least place them in the dishwasher.

3.2 Misconceptions about the wisdom level

If one is not aware of the concept of the data pyramid, there are two quite common misconceptions. They can be observed both in academia and industry. In a somewhat exaggerated form, they can be formulated as

- Data *is* wisdom and
- Visualization of data *is* wisdom.

Where do these—according to the data pyramid—apparently oversimplified views originate from? *Heuristics* are efficient mental processes that help humans solve problems and learn new concepts. They can be seen as mental shortcuts or rules of thumb for finding best-guess answers to existing real-world problems. In effect, the associated processes make problems less complex by ignoring some of the information that's coming into the brain, either consciously or unconsciously. Today, heuristics have become an influential concept in the areas of judgement and decision-making.[2][2]

Concerning the *data is wisdom* misconception:

The problem is probably linked to how we—as natural scientists—typically obtain data from experimental devices, machines or any other, probably even custom-built, experimental setup. Most of the time, the tools merely provide data in a more or less accessible way. The effort of "getting hands on" this data, i. e., preparing them for visualizing or carrying out any other processing step, *feels* like a huge step. In reality, we did not move up to the next level of the pyramid. We have just made the first step towards the interface between *data* and *information*.

Concerning the *visualization of data is wisdom* misconception:

There is a similar issue related to the visualization of data. In order to be able to visualize—in whatever way collected—data we first have to make it accessible to the visualization step (and hopefully do not suffer from the first misconception). However, the next pitfall is just around the corner: Visualization itself is quite an effortful business and includes some mental footwork. The challenge is described by the fact that the abstraction of the real world leads to data, which is encoded in shapes and colours.

2 https://www.thoughtco.com/heuristics-psychology-4171769

This *is* visualization. A good visualization allows a reader or spectator of a visualization to readily decode it. He or she will get a sense of the underlying data and obtain a picture of the real world [8]. This back and forth between visualization of data describing the real world and decoding the visualization to obtain an impression of the real world is far from trivial. The interplay between visualization and interpretation is sketched in Figure 3.3.

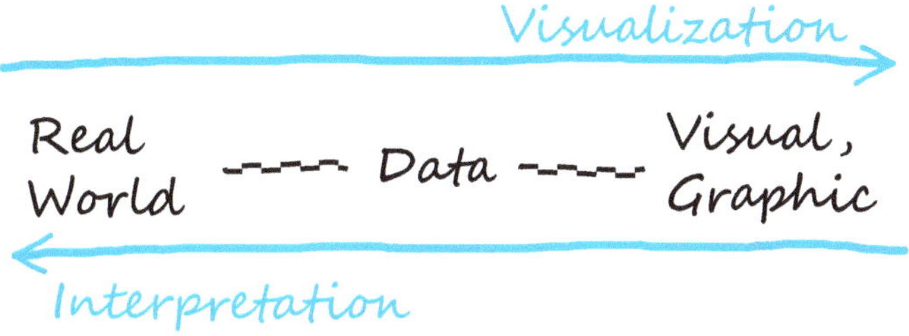

Figure 3.3: The interplay between visualization and interpretation. *Data* (or *information*) serves as the connection between the real world and visualization.

In that sense, data can be considered the "glue" connecting the real world, as we know, experience and—as natural scientists most obviously—measure it, and its visual representation. Despite all the thinking associated with generating a probably convincing, elegant, on-point or even artesian visualization of data, we remain where we used to be—on the level of *data*.

As indicated in Figure 3.1, the scope of this book extends primarily to the lower levels *data* and *information*. In particular, the book deals with

- the identification and extraction of relevant *data* from experimental raw data files (see Chapter 4) and
- the extraction of meaningful *information* from the obtained clean *data* relying on expert knowledge or domain knowledge (see Chapter 5).

3.3 Navigating through the levels of the data pyramid

Next to *data* being visualized, also *information* is subject to visualization. In the context of this book and for scientific results, *information* refers to something like peak positions, characteristic kinks or bumps in series of data points, maximum temperatures in the course of a chemical reaction, average volumes delivered via a pump within a certain time frame or anything else you can think of. A schematic representation of *data visualization* compared to *information visualization* is given in Figure 3.4.

In this admittedly simplistic example, the flow of a medium being pumped by two different pumps A and B powered at variable voltages is represented graphically on the left-hand side. We see immediately from the visualization of the *data* that pump B

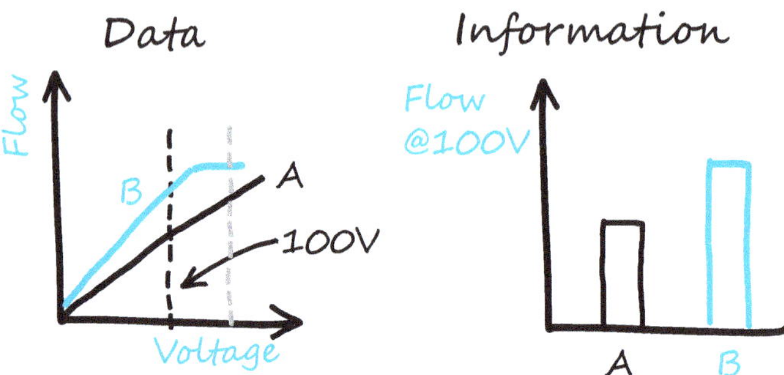

Figure 3.4: Visualization of *data* and *information* compared. The transition from *data* to *information* is associated with a reduction, aggregation or any other "boiling-down" process (see Chapter 5).

delivers greater flow at each operating voltage compared to pump A. Now, lets assume that we use a voltage of 100 V in our setup. The (relevant) *information* we extract from the *data* visualization is obviously the flow associated with both pump options A and B at this operating voltage. A possible visualization of this *information* is a vertical bar plot of the flows at the selected voltage for both pumps. This step from *data* to *information* comes along with two important realizations:

- It is associated with a reduction in the number of points to be represented. We reduced the apparently large number of flow-voltage pairs represented by the line in the left-hand graph to a single value represented by the heights of the bar plots in the right-hand graph of Figure 3.4.
- Only the addition of context, i. e., the knowledge that we want to operate the pump at 100 V, makes possible taking the step from *data* to *information.*

Adding further meaning to this information, we see that, under those conditions, pump B delivers approximately double the flow of pump A. From another perspective, pump B delivers the same volume as pump A in roughly half the time. This can be understood as the *knowledge* level. Until this stage, everything was directed to observations from the past.

If we accept that the *wisdom* stage is related to future situations, *what ifs* and imagined alternative scenarios, we have to draw insights from the knowledge level. For example, we are faced with the challenge of pumping a higher volume of a certain liquid into a reactor because the pumped raw material is, for whatever reason, available only in a much less concentrated quality. However, as a further boundary condition, the pumping step may take only additional 15 min for processual reasons. What is the minimum concentration of the reactant you could process using pump A? Under which conditions would you have to switch to pump B?

But there could be even more challenging questions making it worthwhile to move seamlessly across the levels of the data pyramid, i. e., from *data* to *information* to *knowledge* to *wisdom* and reverse. In each step, you make some assumptions, of course, to the best of your *current* knowledge. It is, however, quite usual that certain aspects and also your evaluations of what is critical and what is not, change in the course of time. And most often, no one is to blame for this.

In our example, let's assume that suddenly there is a stronger power supply available that operates at 110 V instead of the previously used 100 V. With this changed context at hand, we can go back to the *data* level and infer the *information*, that in this scenario, namely, the difference between pumps A and B is less pronounced than before (see values at the intersection points of curves corresponding to pumps A and B with the light grey vertical line indicating the available power suply of 110 V assuming the new scenario in Figure 3.4). As a final consequence, we might end up with the actionable conclusion that each of the pumps considered up to this point might not be appropriate to deliver especially low concentration material into a reaction vessel within a constrained time frame. So, it is time to search for an alternative pump C or to check further options for adjusting the process by other means.

3.4 Wrap up

This chapter introduced the *data pyramid* with its "backward-looking" *data*, *information* and *knowledge*, and "forward-looking" *wisdom* levels. Two common misconceptions related to our perception of *data*, their visualization and their relationship to the *wisdom* level were highlighted. The chapter concludes with an example showcasing the need for seamlessly navigating between the levels of the data pyramid and potential practical implications.

4 From experimental files to data

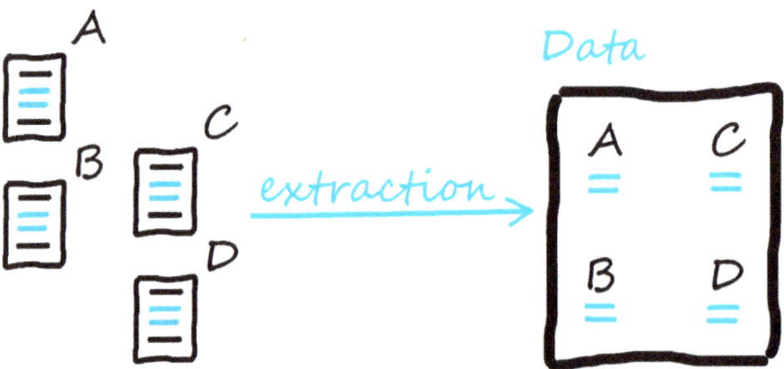

For you as a natural scientist, getting the data is bread-and-butter business. In most typical scenarios, you go to the lab, select your preferred instrument, prepare your sample and carry out the actual measurement. Depending on the technique used, the actual process of measurement is done within seconds, minutes, hours or even days. No matter what, in the end you will be facing a more or less well-structured file of experimental results in almost every case. Many instruments are connected to a "network" that enables direct access to files of experimental results from your working machine. In case of a lower degree of integration at your laboratory, you will need to rely on a universal serial bus (USB) stick to transfer the results file from the measurement machine to its *final destination*, i. e., the location where processing, analysis and/or visualization takes place. In an ideal world, you will be able to store the results file in the preferred place from the very the beginning.

4.1 Challenges

There are some short remarks to be made about experimental result files as they are regularly obtained from the machines of large instrument suppliers:

- They mostly have a particular structure that allows interfacing with the proprietary software packages provided by the respective supplier along with the machine. This is the setup for a commonly occurring problem: Assume that you have an old machine from supplier A to measure a certain quantity of interest but need to replace this device due to ever-increasing maintenance costs and the inaccuracy of the obtained results. You are also aware that in the meantime supplier B has appeared on the market with a far superior device to measure exactly the same quantity at higher precision and lower cost. Would you select the device from manufacturer A or B if you know that it will be quite cumbersome to put results from the different vendors side by side?
- Another issue related to exported results from a machine is that, even if you stick with supplier A for the measurement of the quantity of interest, there is no full guarantee that the format of the result files will *always* be consistent for the fore-

https://doi.org/10.1515/9783110788433-004

seeable future. Some minor firmware updates of the device might be sufficient to lead to difficulties in comparing results obtained over a longer time scale, say a few years.
- Another—interestingly quite common—undesired behaviour of machine exports is the deviation in formats of *single* and *multiple* exports. This means that the file structure you obtain when exporting the results corresponding to an individual sample might look different from an export file holding data on two or more samples.

You definitely should invest some time to verify if any of these issues apply to the machine you are using before diving into the scientific aspects of your work and setting up larger series of experiments to be carried out.

Check and familiarize yourself with the format of export files before measuring larger quantities of samples. Run dummy experiments using multiple experimental settings that might be relevant for you or your team. Does this have an influence on the form of the exported file format? If so, what changes?

4.2 Introducing the exemplary results file structure

I am well aware that these ideas are rather abstract. Therefore, I would like to introduce a—hopefully—readily comprehensible example. As this will be used for the remainder of the book, I would like to supply you with a basic understanding of this quantity. In the following, we will look at so-called *surface tension isotherms*. Anybody in the fields of chemistry, material science, physics, medicine or the like, probably has heard about it at some point in his or her career.

Surface tension is the property that enables a liquid to resist an external force. A quite popular demonstration is the “overfilling” of a glass of water. You are able to fill a glass slightly above the level of its upper rim without spilling water. Likewise, you should be able to gently place a paper clip on top of the surface without sinking it—at least with some skill after a few tries. In this situation, the surface tension is sufficiently large to support both the water level above the rim and the paper clip.

Now, consider the following modification of the setup: You just add a tiny drop of dishwashing agent onto the surface. The result: The water spills and the paper clip sinks as indicated in Figure 4.1.

So, what has this small experiment to do with the previously mentioned surface tension isotherms, which will be used as an example? The glass-water-dishwashing agent setup can be understood as the minimum implementation of a surface tension isotherm experiment. In the laboratory world, you would typically measure the surface tension[1] of an aqueous solution at both a defined concentration and defined temperature. In our simplified version of the experiment, we just evaluate surface ten-

1 There are purpose-built machines to precisely measure surface tension. They are called tensiometers.

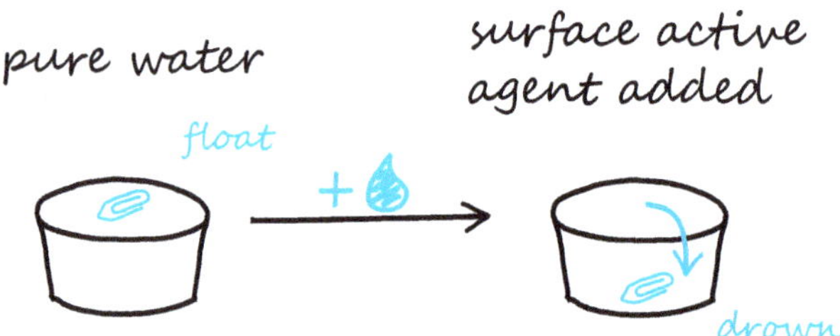

Figure 4.1: The influence of a surface active agent (surfactant) exemplified. Whereas the large surface tension of pure water supports the weight of a paper clip to leave it floating, the addition of a small amount of surfactant changes the picture. The induced reduction in surface tension causes the paper clip to sink.

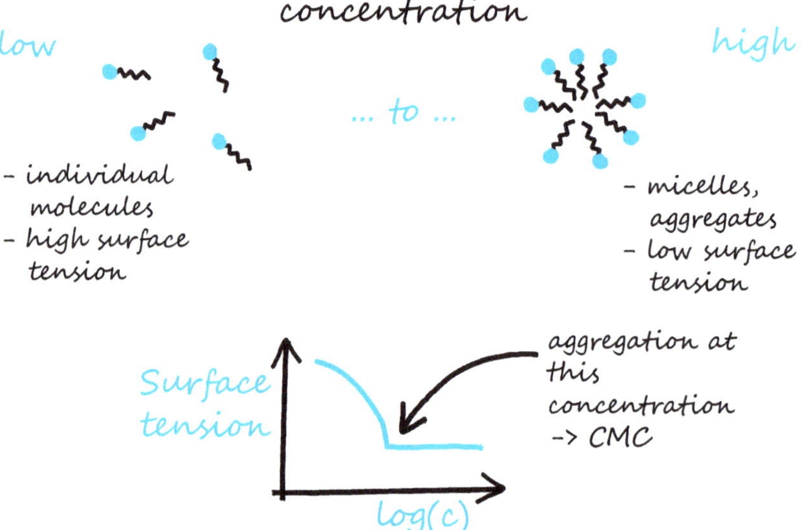

Figure 4.2: Introduction of the concept of a critical micellar concentration. At low concentrations, surfactant molecules are more or less independent from one another and lead to a further reduction in surface tension upon increasing concentration. At high concentrations, the molecules have a tendency to aggregate, thereby forming so-called *micelles*. The CMC marks the concentration above which micelles are formed and is considered a characteristic parameter of the surfactant on the *information* level of a surfactant solution. Typically, the surface tension plot is shown on a logarithmic concentration scale.

sion in terms of large (paper clip floats) or small (paper clip sinks) in the absence or presence of a surface active agent or so-called surfactant, also known as (a. k. a.) dishwashing solution at room temperature.

If we extend this basic experiment and carry out more precise measurements of surface tension instead of the results “paper clip floats” or “paper clip sinks” at multiple concentrations and constant temperature, we end up with the surface tension isotherm sketched in Figure 4.2. The parameter (a quantity on the *information* level) we seek to extract in the following chapter is the so-called critical micellar concentration (CMC). It marks the concentration above which no further decrease in surface tension upon increasing surfactant concentration is observed. On a molecular level, the surfactant molecules tend to aggregate instead of moving to the interface.

Relying on more advanced equipment than usually available in any home kitchen, you might receive a file of the following exemplary format from a dedicated machine as shown in Listing 4.1.

Listing 4.1: raw_from_machine/Polysorbate40.csv

```
=== GENERAL INFORMATION ===
Sample: Polysorbate40
Measurement performed on: 2022-01-28 13:37:44
Operator: Matthias Hofmann
Device: Fancy Machine X
Device ID: Y0139836

=== COLLECTED DATA ===
Concentration / [g/L],Surface tension / [mN/m]
4.0790e-01,43.78
1.4911e-01,43.84
8.8993e-02,43.78
3.9720e-02,46.68
1.9793e-02,49.43
9.2199e-03,53.69
2.7936e-03,57.40

=== SUMMARY INFORMATION ===
7 concentration - surface tension value pairs collected.
Thank you for using Fancy Machine X.
```

The next step is to understand the exported file's structure and answer the following questions:
- What is useful?
- What can be discarded?
- What is missing?

Clearly, answering these questions is not always as easy as in this case and often requires profound domain knowledge. So, at this point, let's recapitulate the actual scientific question we had: What is the surface tension isotherm and the CMC of the studied aqueous surfactant.

The most essential part of the experiment is at the centre of the file shown in Listing 4.1: The concentration–surface tension pairs making up the isotherm. One thing that has to be clarified is whether you are fine with the units of the experimental quantities therein. You can use either conventional units that typically make the numbers easier to grasp or SI-units. There's beauty in either approach, but this will not be discussed here in more detail. Just one thing: Use one or the other consistently—transformation for the purpose of visualization at a later stage should not be the main criterion for the decision.

A part that is obviously to be discarded in the exemplary file shown in Listing 4.1 is the "summary information" part in the bottom lines. This does not add any information to the core part of the collected values nor to a better understanding of the experimental procedure itself.

The "general information" block on top of the file indeed holds some valuable pieces of information. Sample name, date of measurement, operator, the device and its ID are certainly things to keep track of for future reference. A typical question related to this type of information would be: *Can we actually compare those two samples? Were they measured under comparable conditions?* This kind of *information* is also referred to as *metadata*.

Metadata is "data that provides information about other data".[2] Probably the most relevant type of metadata in the scientific context is *descriptive metadata*. It may comprise information on the measured sample, instrument settings, environmental conditions, references to a certain project or scientific target intended with the respective measurement.

But is the information contained in the file really enough? What might be missing? In the realm of measurement terminology, there is the word *isotherm*, meaning that the experiment should be conducted at constant temperature for all the studied concentration–surface tension pairs. So far, there is no information on the, e. g., mean temperature during the experiment. This is something to keep track of in our documentation. The same holds for the medium in which the surfactant is dissolved. Typically, this is water, but alternatives are certainly possible. For example, you might be interested in the surface tension isotherm of a pure surfactant in some type of salt solution or characteristic reference fluid specific to a certain discipline. Therefore, we should document that the experiments in the present example were carried out in water.

Moving on to the sample itself. Is the name as specified in the file really everything we need to unambiguously identify our sample? Probably not. Again, there are multiple ways to approach this identification task in greater detail.

From the chemical side, there is a so-called CAS-number. It is maintained and assigned by the Chemical Abstracts Service (CAS) to uniquely identify a chemical substance.[3] With this number at hand, you have a loose end to further material-specific

2 https://en.wikipedia.org/wiki/Metadata

3 https://www.cas.org/support/documentation/chemical-substances/faqs

information such as molecular mass, chemical structure, safety relevant information and many more associated with that number. If you are interested in these properties, just go for the CAS-number of the sample.

From the accounting or bookkeeping perspective, there might also be some further interesting points to add in order to help identify the sample. For example, your company or institute might receive samples with the same chemical identity, i. e., same CAS-number from different suppliers or materials from several productions of one and the same supplier. To uniquely identify material from one shipment, some type of order identification number might be the thing you are looking for. Having this information at hand, you will be able to search for information associated with this very material in the ordering system. The order number will serve as the loose end to information such as the manufacturer, production date, cost, ordered quantity, shipment date and many others.

At this stage, physically printing an example results file (if not too large, of course) and enriching it with some of the information just listed might be a helpful step in your data processing routine. An example introducing the general idea is shown in Figure 4.3

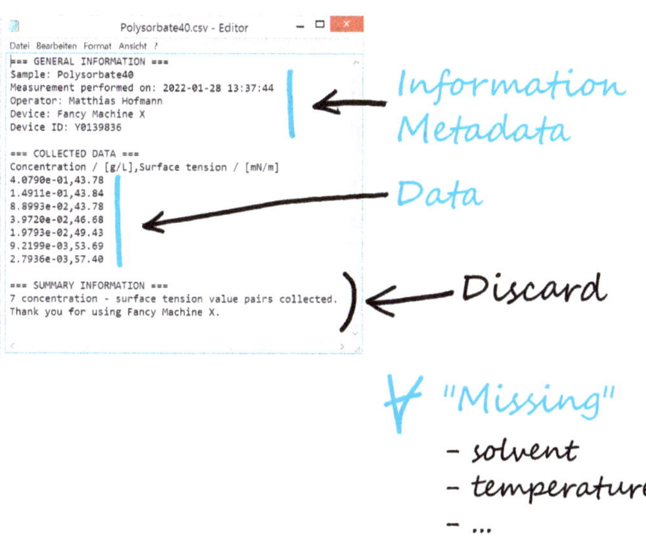

Figure 4.3: Screenshot of an opened experimental results file with highlighted parts for the identification of relevant parts such as *data* and *information* (here *metadata*) according to the DIKW model. In this context, also *missing data* and *missing information* should be thought of. Alternatively, a physical printout can be used for this purpose.

A physical printout of an exemplary results file is helpful for structuring your thoughts on relevant, irrelevant and missing pieces of information. i

---

So, how can we make the step from the shown results file in Listing 4.1 to the *data* (and *information* in the sense of the previously introduced *metadata*) that we are actually interested in? To sum up, in our example those are
- the concentration–surface tension pairs,
- information on the measurement (such as the experimental temperature) and
- information on the sample.

As previously stated, the *information* part has to be enriched by further entries that are *not* available in the file originally obtained from the machine.

4.3 Pseudo-code

Writing some pseudo-code before actually diving into the scripting work is recommended and good practice—especially for beginners. The main benefits are
- obtaining a rather abstract view of the desired processing steps (What do I want to achieve?),
- having access to a high-level overview of the involved steps, both for later reference and not losing sight of the key objective of your piece of code, and
- having a starting point for more detailed comments within the script.

Without any further knowledge about the syntax, the pseudo-code for the step *From Experimental Files to Data* might look as shown in Listing 4.2:

Listing 4.2: pseude_code_files_to_data.py

```
#
# 1) Read experimental file content

#
# 2) Extract "Data" and "Parameters" (Metadata) from file content

#
# 3) Maybe? processing of "Data" and "Parameters"

#
# 4) Maybe? add additional pieces of information to "Data" and
#     "Parameters"

#
# 5) Save "Data" and "Parameters" to files or database
```

4.4 Shortcut for accessing data and information: pandas-functions

In the case of a rather "accessible" results file structure as shown in Listing 4.1, carrying out the steps described in Listing 4.2 is drastically simplified by using the pandas library. This approach, which is potentially less stable in terms minor variations of the files' structure, is fair enough in many cases, but we will not focus on this in the present discussion. The code for extracting *data* and metadata, i. e., *information* from exemplary file using *pandas*, is shown in Listing 4.3.

Listing 4.3: extract_from_raw_file_pandas.py

```
# pandas
import pandas as pd

# define filename
file = "Polysorbate40.csv"

# get data
data = pd.read_csv(
            file,
            header=7,  # get column names from line number 8
            skipfooter=3,  # discard bottom 3 lines
            names=[
                "concentration_g_l",
                "surface_tension_mN_m"
                ],
            engine="python"
            )
# print data
print("Collected DATA:")
print(data)

# get "metadata" (information level)
metadata = pd.read_csv(
            file,
            skiprows=1,  # skip to line
            nrows=5,  # number of lines holding metadata
            sep=": ",  # column separator between parmeter name and
                       # value
            header=None,  # do not read column names from file
            names=["parameter", "value"],  # specify column names
                                           # explicitly
            engine="python"
            )
# print information (metadata)
print("Collected INFORMATION:")
print(metadata)
```

The console yields the following output:

```
Collected DATA:
   Concentration / [g/L]  Surface tension / [mN/m]
0               0.407900                     43.78
1               0.149110                     43.84
2               0.088993                     43.78
3               0.039720                     46.68
4               0.019793                     49.43
5               0.009220                     53.69
6               0.002794                     57.40
Collected INFORMATION:
                  parameter                value
0                    Sample        Polysorbate40
1  Measurement performed on  2022-01-28 13:37:44
2                  Operator     Matthias Hofmann
3                    Device      Fancy Machine X
4                 Device ID             Y0139836
```

This approach works perfectly fine if you are certain that you are working with *exactly* the same type, i. e., you are sure which information to find at which location within the file. In the following sections, the same *DataFrames* will be obtained from the same experimental raw data file using a more generalized method. If this pandas-based approach is sufficient for the specific use case in mind, feel free to proceed to Chapter 5.

Use the already introduced *pandas*-based method and the approach described in the following, and compare the resulting `DataFrames` `data` and `metadata` obtained from both procedures.

4.5 Reading file contents

Now, let's move to the interesting part of splitting the file according to the distinction just made. Please note that the format of the example files used herein is comma-separated values (csv). This is a basic format that allows for easy reading and handling via Python. A further advantage of *csv*-files is that almost any device provides some plain text or *csv*-export as a fallback option. Use this wisely. Reading the contents of our exemplary results file `Polysorbate40.csv` to the string variable `results` is shown in Listing 4.4.

Listing 4.4: extract_from_raw_file.py

```
# define filename
file = "Polysorbate40.csv"

# initialize results variable of type string (empty)
results = ""

# open the file for reading "r" with encoding utf-8
with open(file, "r", encoding="utf-8") as f:
    # read file line by line
    while True:
        # read line
        line = f.readline()
        # append to string
        results = results + line
        # end of file reached?
        if not line:
            print(type(line))
            # break infinite while loop
            break
```

New elements found within this code snippet are the `open`-function and the `while`-loop.

Arguments of the `open`-function are the path of file to be read, the text mode argument `r` for "reading" and an optional argument to specify the encoding of the file. The latter is a typical pitfall especially when dealing with export files from older devices.

With the opened file, as specified by the indent, we do some reading until its end via the `while`-loop. A *while*-loop runs indefinitely until a certain condition has been met. Because the end of the experimental raw data file is certain, there is no danger of being "captured" in the loop forever. The end of the file is reached when the return value of the file's `readline`-method does not hold any more content. For each of the other previous lines, the string content is appended to the `str`-type variable `results`.

4.6 Regular expressions for accessing data and information

The tool we are going to use to separate the contents of an experimental file as obtained from a machine into the *data* and *information* parts are *regular expressions*. In the programming community, they are referred to as *regex*. A regex is a sequence of characters that specifies a search pattern in text.[4] For example, they enable finding email addresses, zip codes, telephone numbers, etc., in text. So far, so good. But, how

4 https://en.wikipedia.org/wiki/Regular_expression

does this concept help us with our task of extracting parts from an experimental raw data file relevant for our purposes?

Just as the previously mentioned email addresses, zip codes or telephone numbers, there are certain *patterns* in our experimental results file. Looking at the *collected data* block in Listing 4.1, these "data lines" have the format:
- number in exponential format representing the surfactant concentration,
- comma,
- floating point number representing the surface tension value.

So, we *just* need to tell the regex module of Python to look for this pattern, and we're almost there. There are numerous books available detailing the full power of using regex. For the scope of the task at hand, I'll just show the most important key features, sacrificing some detail and possibly elegance. On the other hand, this comes with an increased readability of the regex *patterns*.

In order to set up the most essential patterns to be found in text, you will need to understand the following tokens, i. e., elements of a programming language.

A programming token is the basic component of source code.

For regex, these tokens are attributed to certain groups of functionality, including:
- general tokens such as new line or tab,
- anchors such as beginning and ends of lines,
- meta sequences such as word or digit characters,
- quantifiers or how often does a certain character appear and
- group constructs.

The tokens summarized in Table 4.1 will be used in subsequent content.

As you might expect, building the appropriate regex pattern from scratch requires some–and by that I mean considerable—training, i. e., trial and error. Therefore, a handy tool enabling you to build your regex skills in a fast and iterative manner is https://regex101.com/. This site makes it possible to try out some versions and provides instant highlighting of the captured groups matched according to the specified pattern. The following steps are involved in building the regex pattern for your Python script:
- Copy the full file content of an experimental file to the *teststring* area of https://regex101.com/ (see Figure 4.4),
- set the *regex flavour* to Python,
- write a pattern of each relevant part to the *regular expression* area and
- write your desired pattern down or copy the developed pattern for later use in your Python script.

Table 4.1: Useful regex tokens in Python flavour.

Token	Meaning
\s	Any whitespace character
\S	Any non-whitespace character
\d	Any digit, i. e., 0 to 9
\w	Any word character
\W	Any non-word character
\.	Literal dot, point, .,
.	Any single character
+	one or more of the preceding
*	zero to unlimited number of the preceding
^	start of string
$	end of string
(?P<name>)	capturing group named *name*

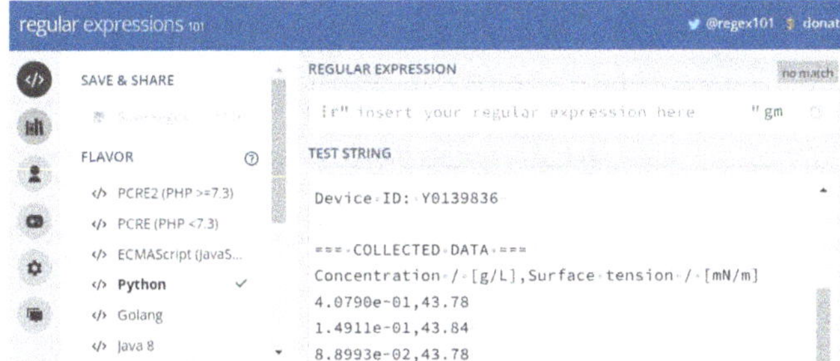

Figure 4.4: Screenshot of https://regex101.com/. On the left-hand side, we select Python as the flavour. The test string, i. e., the content of the experimental raw data file, is copied to the *test string* area at the bottom right. The regular expression to be developed is specified in the identically named area at the top right. Initially, the placeholder "insert your regular expression here" is shown.

To get a better impression of the procedure of finding an appropriate regex, I would like to show a first screenshot of https://regex101.com/ in Figure 4.4.

Keep in mind, that there are typically several more or less *right* solutions to define a pattern. Probably, you should rely on a version that supports understanding what you did there a few weeks or months earlier. So keep it simple.

4.6.1 Accessing data from the *Collected Data* section

Now that you are familiar with the very basics of regex, let's get to work and get the *data* from an available *csv*-file. As previously mentioned, we are looking for a "number in exponential format followed by a comma followed by a floating point decimal". This is a line holding *data*, i. e., a pair consisting of concentration and surface tension. As exponential format numbers consist of digits, a decimal point, the character "e" and a minus sign (see Listing 4.1), a possible first approach might be to use the *any character token* represented by a literal dot. In order to identify many of those characters, we translate this to regex via the * token. If we just enter ".*" in the regular expression area line entry, *everything* is highlighted as shown in Figure 4.5.

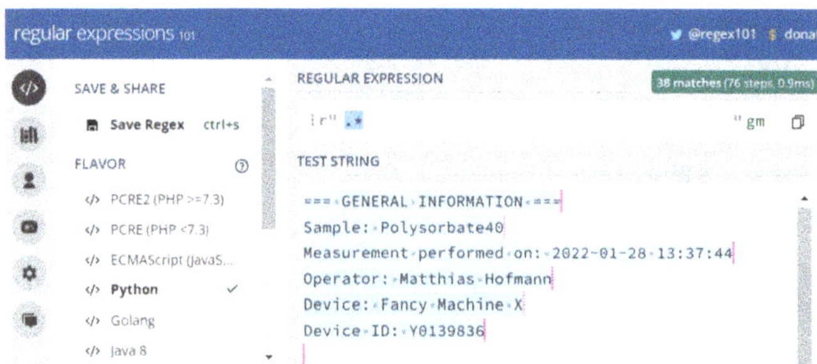

Figure 4.5: We start by allowing *any* single character (.) occurring from zero to an unlimited number of times (*). Accordingly, all lines are completely highlighted.

No reason to panic. Let's just move on. The separator between the columns holding *concentration* and *surface tension* is a literal comma. So, let's append this sign to our regex. The number of highlighted matches significantly reduces as shown in Figure 4.6.

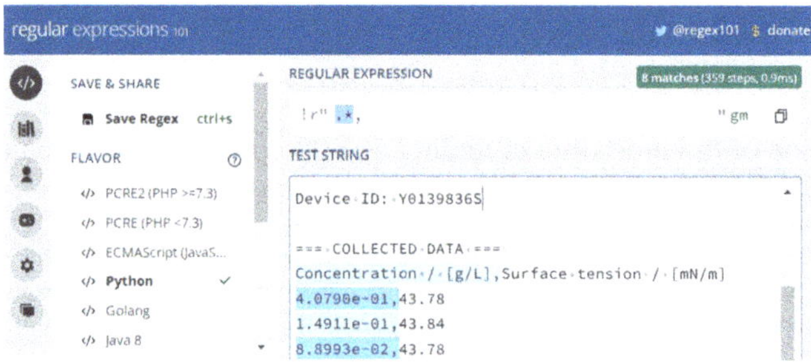

Figure 4.6: Additional specification of the literal comma (,) leads to less findings, i. e., a smaller portion of test string indicated by highlighting. The top three lines are not matched, as there is no comma inside them.

We just need to get rid of the line specifying the column names. Focusing on the first line holding actual data, we see two digits after the comma, so let's translate that into regex as "one or more digits". This will lead to a highlighting of only the lines in the file that we would naturally associate with the experimental *data*. In order to avoid the truncation before the decimal separator, we need to collect the rest of the line by searching for more arbitrary characters. Here, we apply the pattern ".+" (read as "one or more of any single character") as shown in Figure 4.7.

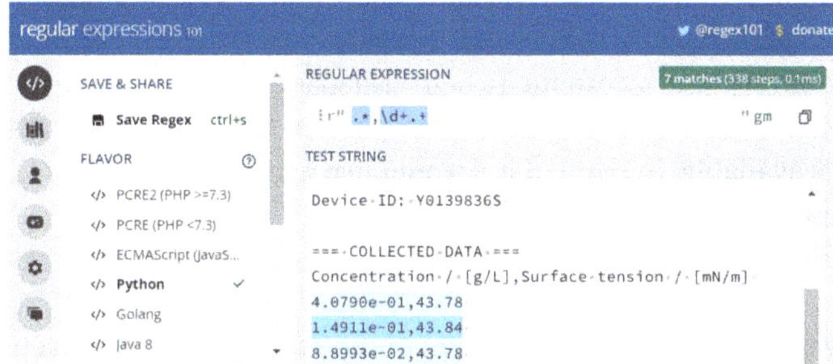

Figure 4.7: To exclusively get hands on the *data* lines, we want one or more (+) digits (\d) to follow the already specified expression. The remainder of the floating point number we are interested in is captured allowing any single character (.) one or more times (+).

So far, we get seven *matches* as indicated on the right-hand side of the window. In order to further simplify the following processing steps in Python, it would be helpful to separate, i. e., distinguish between, concentrations and surface tensions already covered within the developed regex pattern. For this purpose, we just enclose the tokens before and after the literal comma in brackets. This will not only add more colour to the interactive regex builder (see Figure 4.8), but will also help us in the following during *data* extraction.

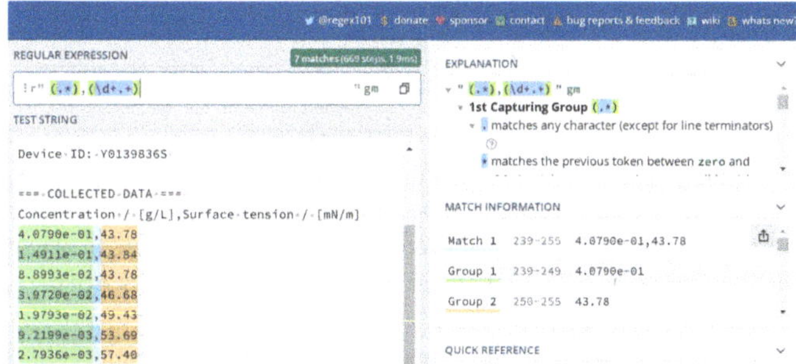

Figure 4.8: For separating the concentration and surface tension values, we further introduce *capturing groups* represented by brackets. In the *match information* section on the right-hand side, the captured groups are highlighted as *Group 1* and *Group 2* within the respective *Match*.

Use each of the files *Polysorbate20.csv* and *Polysorbate40.csv* to develop the regex pattern shown in Figure 4.8. Furthermore, check the validity of the pattern for matching the *data* parts of these exemplary files.

As visually indicated in Figure 4.8, each individual *Match* is subdivided into two groups: the first group representing the concentration and the second group the surface tension.

So let's recapitulate our pattern once more. To help visualize and understand what we have actually done here, and also to do so at a later point in time, there are some regex visualization tools available. In particular, I would like to refer you to

- https://www.debuggex.com/ and
- https://extendsclass.com/regex-tester.html#python.

With the help of those visualizers, it is clearly easier to understand your regex conceptually. Especially if writing regular expressions is not your everyday business, these tools might come in handy. In our example, we defined the regular expression (.*),(\d+.+) to match our *data* lines in the experimental raw data file. Using the graphical representation from *extendsclass.com*, we get a somewhat clearer understanding of the tokens and their grouping as shown in Figure 4.9.

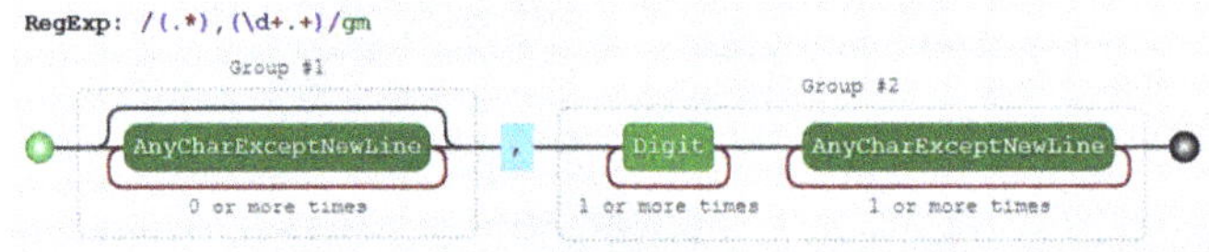

Figure 4.9: Visual representation of the derived regular expression (.*),(\d+.+) from *https://extendsclass.com*. The regex comprises two groups separated by a comma, whereas group #1 consists of any character occurring multiple times, and group #2 consists of multiple digits followed by multiple other characters.

The previously described regex is not very strict. This comes with both advantages and risks. For example, depending on the machine from which experimental raw data files are exported, different decimal separators may appear. Another common case is the precision of floating point numbers in experimental raw data files. It is crucial at this stage to test your regex with several example files as early as possible to account for potentially erroneous matches.

One remark at this point: A regex is appropriate for the extraction of the respective *data* if you can match *exclusively* the block of interest. In our example, other valid regular expressions for the purpose of extracting *data* lines from experimental raw files are:

- (\d\.\d+e-\d+),(\d{2}\.\d{2})
- (\d\..*),([\w\.]+)

Their graphical representations are given in Figure 4.10.

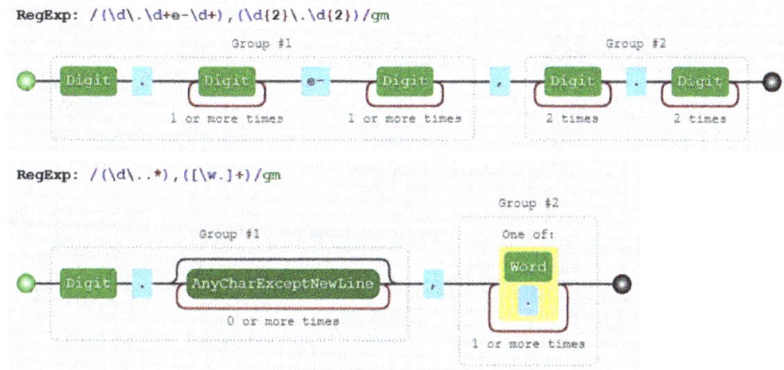

Figure 4.10: Visual representations of the alternative regular expressions from https://extendsclass.com to match a data line in the exemplary experimental raw data file.

In order to get an understanding of the mechanics of regex, try to rebuild the pattern versions shown above and think about possible weaknesses of each.

Now that we know how to extract the *data* part from the initial experimental raw data file, let's get back to our Python script and apply the gained knowledge on regex. In order to leverage regular expressions in Python, we use the `re` package. In particular, we need the `findall`-function of `re`.

As you can easily see from spyder's variable explorer, the `findall`-function returns a list of tuples. Our next step will be to transform this list of tuples into an easily manageable `DataFrame`.

So, here we are again—looping a `list`: In this very example, we want to use the first element of each of the resulting tuples as concentration and the respective second element as the corresponding surface tension.

Python additionally provides a shorthand syntax for creating a new list based on the values of an existing list, i. e., *list comprehension.*[5] Both options shown in Listing 4.5 for extracting quantities of interest are equally valid. Maybe the first version appears more natural to you, so just stick with it. In case you prefer a more concise code, list comprehension might be the way for you to go.

5 https://www.w3schools.com/python/python_lists_comprehension.asp

Listing 4.5: extract_from_raw_file.py (continued)

```
# %% extract data via regex
#

# pattern as derived via "regex101.com"
pattern = "(.*),(\d+.*)"

# import python regex module
import re

# find pattern in string "results"
findings = re.findall(
                pattern,
                results
            )
# show "findings" variable (list of tuples)
print(findings)

# import pandas module to use the DataFrame with alias pd
import pandas as pd

# initialize empty pandas.DataFrame data
data = pd.DataFrame()

# initialize empty list of concentrations
concentration_g_l = []
# loop through findings to get concentrations
for _finding in findings:
    # info
    print(_finding)
    # get the first element of the "_finding" (string) and convert
    # to float
    _c_g_l = float(_finding[0])
    # append extracted concentration to list of concentrations
    concentration_g_l.append(_c_g_l)

# use concentration list as column in the defined DataFrame
data["concentration_g_l"] = concentration_g_l

# use surface tension as colums (via list comprehension)
# type conversion to float via "float()"-function
data["surface_tension_mN_m"] = [float(i[1]) for i in findings]

## alterantively: build DataFrame from list of tuples available from
## "findings"
#data =  pd.DataFrame(
#            findings,  # structured source of data
#            columns=[  # specify column names
```

```
#                    "concentration_g_l",
#                    "surface_tension_mN_m"
#                    ],
#           dtype=float  # force conversion to float
#           )

# print resulting DataFrame
print(data)

# clean variable space
del concentration_g_l, pattern, findings
```

The console yields the following output:

```
[('4.0790e-01', '43.78'), ('1.4911e-01', '43.84'),
('8.8993e-02', '43.78'), ('3.9720e-02', '46.68'),
('1.9793e-02', '49.43'), ('9.2199e-03', '53.69'),
('2.7936e-03', '57.40')]
('4.0790e-01', '43.78')
('1.4911e-01', '43.84')
('8.8993e-02', '43.78')
('3.9720e-02', '46.68')
('1.9793e-02', '49.43')
('9.2199e-03', '53.69')
('2.7936e-03', '57.40')
   concentration_g_l  surface_tension_mN_m
0           0.407900                 43.78
1           0.149110                 43.84
2           0.088993                 43.78
3           0.039720                 46.68
4           0.019793                 49.43
5           0.009220                 53.69
6           0.002794                 57.40
```

Note that del in the Listing 4.5 is a keyword used to delete objects. Just as in Python, everything is an object, and this keyword can be used "clean up" the variable explorer.

4.6.2 Accessing data from the *General Information* section

We can proceed in a similar manner for the extraction of the *general information* section of the raw data file as shown in Listing 4.1. Also here, we need to specify our regex pattern. In contrast to the previously shown example, we will make use of *named capturing groups* for the identification of the respective parameters and values. In our example, the character sequence “colon followed by space” represents the separator between parameter name and value, or *key* and *value* as you like. Accordingly, a possible regex pattern for the extraction of these values is given in Figure 4.11. Additionally, a graphical representation of the pattern is shown in Figure 4.12.

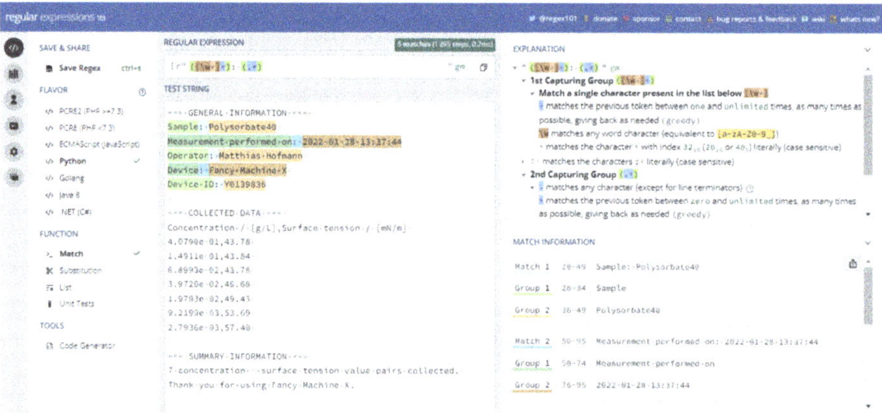

Figure 4.11: Regular expression to match the *information* lines contained in the experimental raw data file. A separation between the information name and its value is indicated by a colon followed by a character space.

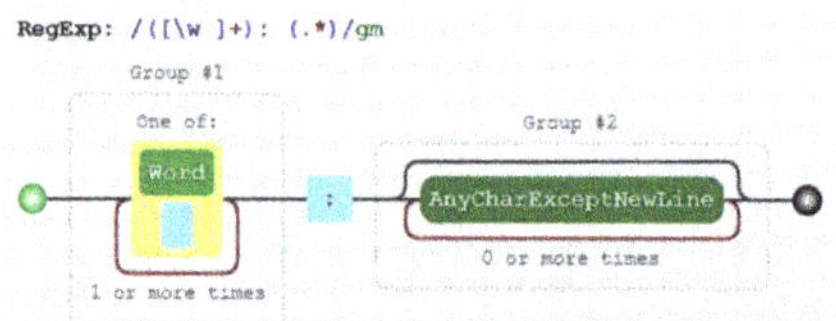

Figure 4.12: Graphical representation of the regex pattern for finding *general information* lines in the experimental raw data file. The latter are on the *information* level of the data pyramid.

In addition to the previous example on “data” extraction, we want to apply naming of capturing groups here. Therefore, we just have to include *?P<name>* at the beginning of the capturing group brackets for naming the group. Overall, we end up with the following regex pattern including named capturing groups: *(?P<parameter>[\w]+): (?P<value>.*)*.

As an alternative to the previously shown function `findall` we will use the `finditer`-function introduced in Listing 4.6. This function returns a *callable_iterator* object. What’s important for using this type of object? We can loop through the `iterable` via the already shown `for`-loop as we did on several previous occasions. Looping over the `iterable` shows us, e. g., via the varibale explorer, that the type of an individual element of the iterable is `re.Match`.

An *iterable* is any Python object capable of returning its members one at a time, permitting it to be iterated over in a for-loop.[6]

So, what's this Match type of the re-module again? It is—especially in the beginning of your coding endeavour—quite common to come across unknown data types and corresponding functionalities. In order to find out more about this data type, we just write re.Match? in spyder's console, and we will be provided with a short description of the data type as shown in Figure 4.13.

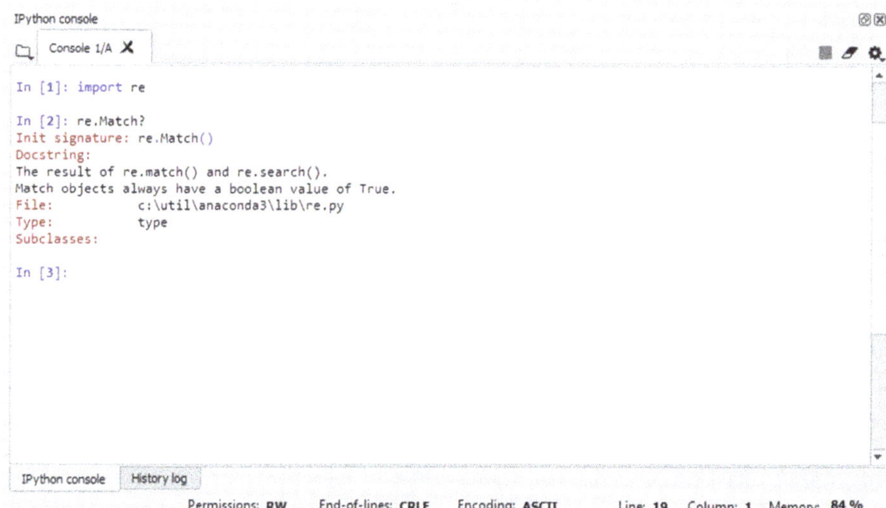
IPython console
Console 1/A
In [1]: import re
In [2]: re.Match?
Init signature: re.Match()
Docstring:
The result of re.match() and re.search().
Match objects always have a boolean value of True.
File: c:\util\anaconda3\lib\re.py
Type: type
Subclasses:
In [3]:
IPython console
History log
Permissions: RW
End-of-lines: CRLF
Encoding: ASCII
Line: 19
Column: 1
Memory: 84 %

Figure 4.13: Screenshot of the iPython console in the spyder integrated development environment (IDE) showing the command line help for the Match type of the re-module.

Typing something you would like to learn about in the console followed by ? will give you some information. This is also possible for your defined variables.

In order to find out what you can actually do with a re.Match object, or in particular with _f as defined implicitly in Listing 4.6 within the for-loop, we will use another helpful tool available in spyder. Typing dir(_f) or dir(re.Match) in the console window will return a list of "functions attached to the object", the so-called *methods*. Going through the list, we find a method called groupdict. Judging from the name, this is something we could potentially make use of for the extraction task at hand. So, let's try this and get some further information via _f.groupdict?. According to the

6 https://www.pythonlikeyoumeanit.com/Module2_EssentialsOfPython/Iterables.html

return of this request (*Return a dictionary containing all the named subgroups of the match, keyed by the subgroup name.*) in Figure 4.14, this sounds just like the thing we are looking for.

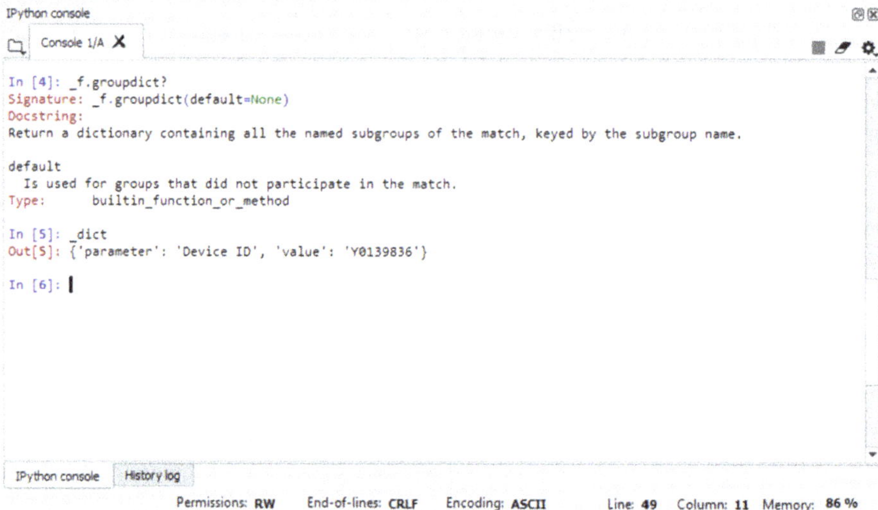

IPython console
Console 1/A
In [4]: _f.groupdict?
Signature: _f.groupdict(default=None)
Docstring:
Return a dictionary containing all the named subgroups of the match, keyed by the subgroup name.
default
Is used for groups that did not participate in the match.
Type: builtin_function_or_method
In [5]: _dict
Out[5]: {'parameter': 'Device ID', 'value': 'Y0139836'}
In [6]:
IPython console
History log
Permissions: RW
End-of-lines: CRLF
Encoding: ASCII
Line: 49
Column: 11
Memory: 86 %

Figure 4.14: Working in the console to get help on the `_f` object's (it is of type `re.Match`) method `groupdict` and the key-value pair of the most recent dictionary `_dict`.

Specifying the name a Python object you would like to learn about within the `dir()`-function will give you some information on `methods` and `attributes` associated with this object.

A *method* is a function that "belongs to" an object. In Python the term *method* is not unique to class instances: Other object types can have methods as well. For example, list objects have methods called append, `insert`, `remove`, `sort`, and so on.

In order to use the parameter names and corresponding values specified via the `dict` obtained from regex search, we have to access the latter through the keys specified in the named capturing groups as shown in Figure 4.14 via the console.

Listing 4.6: extract_from_raw_file.py (continued)

```
# %% extract parameters via regex
#

# define regex pattern
pattern = "(?P<parameter>[\w ]+): (?P<value>.*)"

# find pattern in string "results" as iterator "findings_parameters"
findings_parameters = re.finditer(
```

```
    pattern,
    results
    )

# initialize list of "parameters" and "values"
parameters = []
values = []

# loop over iterable to get each "parameter" and "value"
for _f in findings_parameters:
    # get dict for this finding
    _dict = _f.groupdict()
    # extract "paramater" and "value" and append to list
    parameters.append(_dict["parameter"])
    values.append(_dict["value"])

# build pd.DataFrame "information" from the extracted list
information = pd.DataFrame({
    "parameter" : parameters,
    "value" : values
    })

## alternatively: get "information" DataFrame directly from list
## of tuples
#information =  pd.DataFrame(
# re.findall(pattern, results),  # structured source of data
# columns=[  # specify column names
# "parameter",
# "value"
# ],
# dtype=float  # force conversion to float
# )

# remove variables which are no longer required
del parameters, values, findings_parameters, pattern
```

Where are we now after all this effort? In short, we reduced a somehow formatted experimental raw data file containing some *data*, some *information* and some additional stuff to two `DataFrames`: one holding *data* and the other holding *information* as taken from the original experimental raw data file. With these in hand, it is much easier to continue with the remaining tasks of data processing and storage.

As previously stated, we are now in a position to add further pieces of information to our `information` `DataFrame`. In our practical example, this will be the type of medium or solvent, experimental temperature, time stamp of processing, i. e., the time of running this very script and the name of the operator who ran the script. Some of these will be set manually, others will be retrieved via appropriate functions as shown in Listing 4.7.

Here, defining a custom function for adding an additional row to the existing `information` is a valid option.

Adding additional rows to a `DataFrame` via the append-function is a bit trickier than for a plain Python `list`. It is important to keep in mind that a `DataFrame` has somewhat more structure than a `list`. Accordingly, we are not able to simply "throw" some values at the `DataFrame` and expect Python to sort out the mapping to existing columns according to our needs. Two options for appending named rows to an existing `DataFrame` are shown in Listing 4.7:

- adding the new row as `dict`, and
- adding the new row as `pd.DataFrame`.

In both cases, a mapping between the values to be appended and the existing column names can be achieved. A closer look to Listing 4.7 shows that this assignment is necessarily independent of the sequence of the specified `_par` and `_var` variables in the respective `dicts` and `DataFrames` to be appended. The mapping occurs via column names instead of indexed location.

Listing 4.7: extract_from_raw_file.py (continued)

```


# %% add further -- considered -- relevant information
#

# import modules for adding further pieces of information
import datetime
import os

# define column names as strings
_par = "parameter"
_val = "value"

# solvent medium (add new row via dict)
information = information.append({
    _par : "Solvent medium",
    _val : "water"
    }, ignore_index=True  # increase index "automatically"
    )
# experimental temperature (add new row via pd.DataFrame)
information = information.append(pd.DataFrame({
    _par : "Experimental temperature degree C",
    _val : 23  # temperature of thermostatted room
    }, index=[len(information)+1])  # increase index explicitly
    )
# processing timestamp (add new row via dict)
information = information.append({
```

```
    _val : datetime.datetime.now().replace(microsecond=0), # timestamp
    _par : "File processed on"
    }, ignore_index=True
    )
# operator (add new row via pd.DataFrame)
information = information.append(pd.DataFrame({
    _val : os.getlogin(),  # get logged in user
    _par : "File processed by"
```

Back to the previously stated high-level question of where this extraction of *data* and *information* from an experimental raw data file leveraging the power of regex has taken us so far: You might wonder with good reason, why we did *not* use the powerful `pandas.read_...`-functions.

In particular, pandas' functions `read_excel` and `read_csv` are very helpful for reading *clean* files. Reading the exemplary raw data file via this approach is shown in Chapter E.

As a first reason, I would like to point out the robustness of the extraction of relevant blocks from a Python string. Take a look at the *collected data* block in Listing 4.1. In this case, we have seven "pairs" or tuples of concentration and surface tension. With our—newly gained or long standing—domain knowledge in the field of surface chemistry, we are well aware that a more or less scrupulous research scientist or technician would have chosen a significantly higher number of concentration–surface tension pairs for her study. Both cases will be covered with the regex-based extraction approach previously described (under the condition that experimental data rows still match the corresponding regex pattern). Relying on the appropriate read function from pandas for *csv*-files, `pd.read_csv`, you would need to know the precise line numbers within the file holding the relevant data. Clearly, you do not have this information before actually looking into the file. In coding terms, this would require some kind of pre-reading before performing the actual data extraction via `pd.read_csv`. Practically, we could define a custom "find" function returning the number of lines holding the *data*-part according to the understanding of this book. Those numbers could then be used further for the actual reading task via `pd.read_csv`. The regex approach allows for some more flexibility in terms of experimental freedom and does not tie you or your colleagues to *exactly* the same experimental procedure leading to identically shaped experimental files.

A second reason for using the regex approach is the availability of a *csv*-export option in almost any case—whatever the exact format in terms of content may be. As long as you are able to read the file's content into a string, you are good to go with the previously described approach. The `pd.read_csv`-function relies on a more structured format of the file to be read in order to make full use of this function's capabilities. You will not be able to read formatted data from *any csv*-file without additional effort.

Reading the file contents of the exemplary results file *Polysorbate40.csv* is shown in Chapter E in comparison to the procedure introduced herein. If the `pd.read_csv`-approach is more appealing to you or sufficient for the specific type of files you will be working with, use this approach.

Taking the previously described extraction task one step further, a possible way to go is defining functions performing the tasks of the developed script. Again, a possible piece of pseudo-code is shown in Listing 4.8.

Listing 4.8: pseude_code_files_to_data_via_functions.py

```
#
# 1) Read experimental file content

#
# 2.1) Extract "Data" from file content

#
# 2.2) Extract "Parameters" from file content
```

4.7 Building the custom `file_to_data` module

Each of the tasks just sketched can be carried out by a dedicated function from a custom module to be defined. For the purposes of our task at hand, we will define a Python package that we will use for all the reading and processing steps covered in this book. In order to get things going, we generate a folder named *surface_tension* in the targeted working directory and place an empty Python file called *__init__.py* in it (see Section 2.4.6). Furthermore, we create another Python file therein with the name *file_to_data.py*. This file's task is to hold the functions handling the steps described in Listing 4.8. The structure used herein is given in Figure 4.15.

Figure 4.15: Structure for using functions from the user-defined package `surface_tension` in the script *files_to_data_via_functions.py*. To make use of the data handling functions defined in the module `file_to_data` available to the script, the containing folder has to contain an empty *__init__.py* file.

In short, we will just reuse the code written in the so-far built file *extract_from_raw_file.py* to arrive at a much more condensed and cleaner version of the steps carried out until this point. Therefore, we cut the respective parts from our initial script, pack it into the corresponding function and add some documentation for later reference. The first part of the task, namely, reading the experimental raw data file content as Python string, is achieved via the function `read_file_content_as_string` as defined in Listing 4.9. Here, we need to introduce one parameter in our reading function: the path of the file to be read.

Listing 4.9: surface_tension/file_to_data.py

```
# required imports
import re
import datetime
import os
import pandas as pd


def read_file_content_as_string(filename):
    """
    Reads the file specified via "filename" and returns its content
    as str.

    Parameters
    ----------
    filename : str
        path to experimental surface tension file to be read.

    Returns
    -------
    file content as string.

    """

    # initialize results variable of type string
    results = ""

    # open the file for reading "r" with encoding utf-8
    with open(filename, "r", encoding="utf-8") as f:
        # read file line by line
        while True:
            # read line
            line = f.readline()
            # append to string
            results = results + line
            # end of file reached
            if not line:
                print(type(line))
```

```
                # break infinite while loop
                break

    # return file content as string
    return results
```

With this first function of our `surface_tension` package at hand, we can write the mere reading-from-file part as easily as shown in Listing 4.10.

Listing 4.10: files_to_data_via_functions.py

```
# import custom module from package "surface_tension"
from surface_tension import file_to_data

# specify file name
file = "Polysorbate40.csv"

# 1) Read experimental file content
file_content_string = file_to_data.read_file_content_as_string(file)
```

The next functions to be defined according to the sketch in Listing 4.8 are the extraction of *data* and *information* `DataFrames` from the experimental results string `file_content_string`. The task of further enriching the `information` part of our results described in Listing 4.7 should be carried out by a further dedicated—but simple—function of the kind `DataFrame` in, modified `DataFrame` out.

Having defined these functions in the file *file_to_data.py* (a. k. a. `file_to_data` module) according to Listing 4.11, our `surface_tension` package is so far complete.

Listing 4.11: surface_tension/file_to_data.py (continued)

```
def get_data_from_experimental_string(experimental_string,
                                      show_info=True):
    """
    Extracts "data" from the experimetal results string and returns a
    pd.DataFrame.

    Parameters
    ----------
    string : str
        string holding the experimental file content.

    Returns
    -------
    data : pd.DataFrame
        pd.DataFrame with columns
```

```
            - "concentration_g_l" and
            - "surface_tension_mN_m"

    """

    # patterns as derived via regex101.com
    pattern = "(.*),(\d+.*)"

    # find pattern in string "experimental_string"
    findings = re.findall(
                    pattern,
                    experimental_string
                )

    # import pandas module to use the DataFrame with alias pd
    import pandas as pd

    # initialize empty pandas.DataFrame data
    data = pd.DataFrame()

    # initialize empty list of concentrations
    concentration_g_l = []
    # loop through findings to get concentrations
    for _finding in findings:
        # info
        if show_info:
            print(_finding)
        # get the first element of the "_finding" (string) and
        # convert to float
        _c_g_l = float(_finding[0])
        # append extracted concentration to list of concentrations
        concentration_g_l.append(_c_g_l)

    # use concentration list as column in the defined DataFrame
    data["concentration_g_l"] = concentration_g_l

    # use surface tension as column (via list comprehension)
    data["surface_tension_mN_m"] = [float(i[1]) for i in findings]

    # return pd.DataFrame holding experimental "data"
    return data


def get_information_from_experimental_string(experimental_string,
                                              show_info=True):
    """
    Extracts "information" from the experimetal results string and
    # returns a
    pd.DataFrame.
```

```


    Parameters
    ----------
    experimental_string : str
        string holding the experimental file content.
    show_info : bool, optional
        flag for further information. The default is True.

    Returns
    -------
    information : pd.DataFrame
        pd.DataFrame with columns "parameter" and "value"

    """

    # define regex pattern
    pattern = "(?P<parameter>[\w ]+): (?P<value>.*)"

    # find pattern in string "experimental_string" as iterator
    # "findings_parameters"
    findings_parameters = re.finditer(
        pattern,
        experimental_string
        )

    # initialize list of "parameters" and "values"
    parameters = []
    values = []

    # loop over iterable to get parameter name and value
    for _f in findings_parameters:
        # info
        if show_info:
            print(_f)
        # get dict for this finding
        _dict = _f.groupdict()
        # extract "parameter" and "value" and append to list
        parameters.append(_dict["parameter"])
        values.append(_dict["value"])

    # build pd.DataFrame "information" from the extracted list
    information = pd.DataFrame({
        "parameter" : parameters,
        "value" : values
        })

    # return pd.DataFrame holding experimental "information"
    return information
```

```


def add_further_information(information, medium=None,
                            temperature=None):
    """
    Adds pieces of information to "information" pd.DataFrame and
    returns the modified pd.DataFrame "information".


    Parameters
    ----------
    information : pd.DataFrame
        DataFrame holding information.
    medium : str, optional
        information on sample medium ("solvent phase"). The default
        is None.
    temperature : float|int, optional
        information on temperature during the experiment. The default
        is None.

    Returns
    -------
    information : pd.DataFrame
        Enriched DataFrame holding additional pieces of information.

    """

    # define columns
    _par = "parameter"
    _val = "value"

    # solvent medium (add new row dict)
    if medium:
        information = information.append({
            _par : "Solvent medium",
            _val : medium
            }, ignore_index=True
            )
    if temperature:
        # experimental temperature (add new row pd.DataFrame)
        information = information.append(pd.DataFrame({
            _par : "Experimental temperature degree C]",
            _val : float(temperature) # temperature of the room/lab
            }, index=[len(information)+1])
            )
    # processing timestamp (add new row dict)
    information = information.append({
        _val : datetime.datetime.now().\
            replace(microsecond=0),  # timestamp
```

```
            _par : "File processed on"
            }, ignore_index=True
            )
        # operator (add new row pd.DataFrame)
        information = information.append(pd.DataFrame({
            _val : os.getlogin(),  # get logged in user
            _par : "File processed by"
            }, index=[len(information)+1])
            )

        # return modified ("enriched") information pd.DataFrame
        return information
```

4.8 Data and information extraction via a custom module

Making use of the functions defined in our `surface_tension` package, the first piece of our data handling and processing journey, i. e., reading the raw file and extracting *data* and *information*, is done within a few lines of code as shown in Listing 4.12.

Listing 4.12: files_to_data_via_functions

```
# import custom module from package "surface_tension"
from surface_tension import file_to_data

# specify file name
file = "Polysorbate40.csv"

# 1) Read experimental file content
file_content_string = file_to_data.read_file_content_as_string(file)

# 2.1) Extract "data" from file content
data = file_to_data.get_data_from_experimental_string(
            file_content_string,
            show_info=False
            )

# 2.2) Extract "parameters" from file content
information = file_to_data.get_information_from_experimental_string(
            file_content_string,
            show_info=False
            )

# 2.3) Add further "known" parameters not captured in the results
# file
information = file_to_data.add_further_information(
            information,
            medium="water",
```

```
                temperature=23
                )

# remove no longer required variables from the script
del file, file_content_string
```

To increase readability even further, we could define a summarizing function subsequently calling the four functions of our `file_to_data` module within the `surface_tension` package with just two return values: the `data` and `information` `DataFrames`. The downside of this approach might be the increased complexity due to the additional layer—or better hierarchy—of functions.

4.9 Further considerations

At this stage, we already have gone a decent part of the way. We extracted the `data` and `information` parts from an experimental raw data file via dedicated user defined functions to yield two dataframes we can easily work with: one holding the *data*, and the other one holding the *information*. The latter has been enriched by some further bits and pieces concerning the processing of the sample and the experimental procedure *not* contained in the original results file.

Despite manufacturers continually trying to include more fields for documenting lots of possibly relevant parameters, there is no guarantee that everything is available just as you need it. In that situation, you need to identify ways to account for that. !

Given the *data* and *information* so far, there are two general questions you can ask:

- What else can I get from my *data*? Are there any more parameters of interest inside the data waiting to be extracted? To phrase this a little bit offhandedly, *data* will not give you parameters, i. e., *information* voluntarily. You have to explicitly ask for it via thorough analysis, such as peak picking, extrapolation, fitting to model functions and the like. The selection of plausible or even best-to-knowledge parameters is subject to the choice of (to be) experts with certain domain knowledge. As a result, you will end up with a richer set of *information* compared to the present state.
- Where do I put the extracted *data* and *information* to? Once you have the clean `DataFrames` discussed so far, it is a valid question to ask where to go from there. In the scope of this book, two possible approaches will be described in more detail:
  - A simple version relying on using the file system or folder structure of your device. Here, files are written and shifted inside certain directories exclusively via Python. This of course implies the need for some discipline on the side of the user.

- A more advanced approach relies on a (local) SQLite-database managed via Python. This is inherently more structured and less error-prone compared to the file-system approach, however, somewhat more challenging to set up and maintain in case of changes required at a later stage.

Please note that parameters—or better "characteristic parameters"—are on the *information* level of the data pyramid. The terms *parameters* and *information* will be used interchangeably from here on.

Depending on what appears to be more natural, you can proceed either with Chapter 5 describing the *Data to Information Step* or Chapter 6 addressing the question of *Where to put Data and Information*, respectively.

4.10 Wrap up

This chapter introduced some of the challenges associated with experimental raw data files. Next, the exemplary results file structure serving as the basis for the considerations throughout this book was introduced. In the following, the key ideas of the *Experimental File to Data* step were laid out in some pseudo-code. The process started with reading the results file as a string. For the purpose of *data* and *information* extraction, the concept of regular expressions was introduced. Thereafter, a custom `file_to_data` module within the user defined `surface_tension` package was created for accessing the previously defined functions in a comfortable way. As a showcase, the simplified and more concise *data* and *information* extraction of a raw data file relying on the custom `surface_tension` package was demonstrated.

The chapter concludes with a reference to possible next steps: extraction of meaningful *parameters* from *data or* storing the so-far obtained `data` and `information` `DataFrames` appropriately. In practice, each of the steps is required in an actual project.

5 From data to information

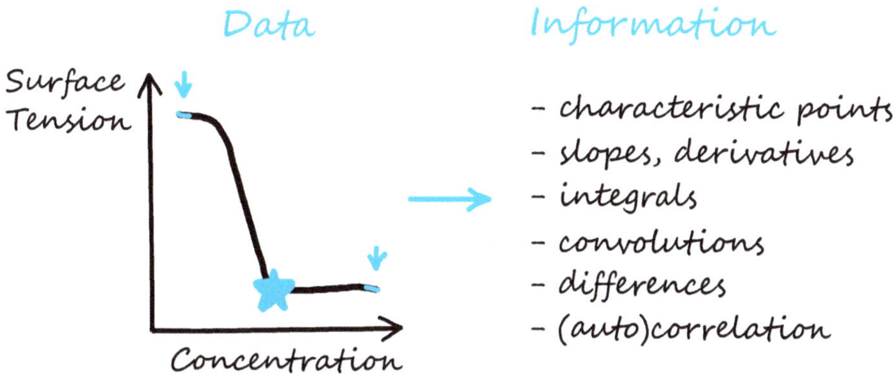

The part of the data processing routine described in this chapter is the actual *fun part* for natural scientists. Here, experts in the respective fields with domain knowledge can make guesses which characteristics of the collected data might be relevant based on their particular scientific interests. In this chapter, I will mostly refer to *parameters*. In this context, parameters are characteristic values on the *information* level of the data pyramid (see Chapter 3).

Roughly outlined, the following *three step approach* will be used to get hands on and store a set of *parameters* corresponding to a sample:

- visualize the collected data,
- extract parameters generally accepted in the field (state-of-the-art parameters) and/or
  extract parameters that might be interesting for the purpose of a certain analysis but that are not regularly considered in the field, and
- compile a set of parameters corresponding to a sample.

5.1 Suggested folder organization

Before actually diving into the analysis, i. e., the parameter extraction part, we need to set up our folder structure appropriately. From what we know so far, we need

- the folder containing the raw data as obtained from the machine (here in the folder *raw_from_machine*) and
- the folder *surface_tension* representing the *package* of the same name developed in Chapter 4.

Schematically, the folder structure used for the extraction of parameters, i. e., *information* from the *data* level, is represented in Figure 5.1.

https://doi.org/10.1515/9783110788433-005

data_to_information_via_function.py
surface_tension
└ _init_.py
└ file_to_data.py
└ data_to_information.py
raw_from_machine
└ Polysorbate20.csv
└ Polysorbate40.csv

Figure 5.1: Folder structure used for the extraction of parameters, i. e., *information* from *data*. The `surface_tension` package defines the functions we need for processing experimental raw data files contained in the folder *raw_from_machine*. In particular, this chapter covers creating the hitherto non-existent module *data_to_information*, which enables, among others, to extract critical micellar concentration (CMC)-values from the so-far collected *data*.

Other than in the previously shown data-reading procedures (see Chapter 4), the Python script itself and the experimental raw data files are no longer located on the same folder level. The raw data files were transferred to the subfolder *raw_from_machine*. This buys us some more organization at the cost of a few lines of additional code for locating these files within the folder structure. For this purpose, we will rely on the `os` module. A suggested first step to ensure proper locating of the raw data files is printing a list of available experimental raw data files according to Listing 5.1.

Listing 5.1: data_to_information.py

```
# import for folder management
import os

# get current directory
current_dir = os.getcwd()
# derive path to data starting from the current directory of the
# script
path_to_data = current_dir + os.sep + "raw_from_machine"

# loop path to data to show list of files
for file in os.listdir(path_to_data):
    # show filename
    print(file)
    # get full filename including path
    file_path = path_to_data + os.sep + file
```

The console yields the following output:

```
Polysorbate20.csv
Polysorbate40.csv
Polysorbate60.csv
Polysorbate80.csv
Polysorbate85.csv
```

An additional step taken therein is the definition of the new variable `file_path` representing the full path to an experimental raw data file. At the end of the shown loop, the variable will persist and can be used in the following for our next steps.

5.2 Pseudo-code

Also, in this situation it is recommended to make use of pseudo-code in order to sketch the intended process in a rather general way. In concrete terms, the steps might be the ones presented in Listing 5.2.

Keep in mind that pseudo-code outlined before going through the analysis step by step is merely a first best guess. It is quite likely that you might find some further interesting aspects about your *data* that might lead to extending the ideas sketched in the initial pseudo-code and the actual code. It is an iterative process. Also, getting advice from, or exchanging with, colleagues is helpful to reduce the number of required iterations.

Listing 5.2: pseudo_code_data_to_information.py

```
#
# 1) Read data

#
# 2) Visualize data --> how? try multiple options as we go

#
# 3) Get "state of the art" parameters

#
# 4) Get further characteristic parameters (time for following
#    hunches ...)

#
# 5) Optional: fit theoretical model to data

#
# 6) "Pack" collected parameters and associate it with the sample
```

5.3 Building on already existing features

Obviously, reading the *data* from our experimental results file is the first step as described in Listing 5.3.

Listing 5.3: data_to_information.py (continued)

```
# %% 1) Read Experimental File content, i.e. "data"

# import custom module from "surface_tension" package
from surface_tension import file_to_data

# get file content as string
file_content_string = file_to_data.read_file_content_as_string(
                                                         file_path
                                                         )

# extract "data" from file content
data = file_to_data.get_data_from_experimental_string(
            file_content_string,
            show_info=False
            )

# remove no longer required variables
del file_path, file_content_string
```

At this point, already something interesting is to be noted. Even though we changed the relative hierarchy of experimental raw data files and the running Python script (remember introducing the subfolder *raw_from_machine* holding the raw data as shown in Figure 5.1), we still can make use of the functions defined in our `file_to_data` module contained in the `surface_tension` package. The `read_file_content_as_string`-function can deal with an absolute path specification, which we can easily provide using the `os` module. The other relevant function, `get_data_from_experimental_string`, does its job just as previously shown. Its only mandatory argument is the file content of the experimental raw data file as a string. No path information is needed for this step. Consequently, there is nothing to be adapted in this function.

5.4 Basic data visualization to identify parameters

The next step on our agenda is a basic visualization. For this purpose, we will use the `matplotlib` library. According to its website, it is a comprehensive library for creating static, animated and interactive visualizations in Python. “It makes easy things easy and hard things possible.”[1] In particular, the rich example gallery is a useful source of

1 https://matplotlib.org/

information for almost any kind of plot you can think of in the context of visualizing natural scientific data. In our case, we will plot surface tension values against the corresponding concentrations, i. e., surface tension on the y-axis and concentration on the x-axis. Using the basic `plot`-function of the `matplotlib.pyplot`-module without specifying anything else according to Listing 5.4, we end up with Figure 5.2.

Listing 5.4: data_to_information.py (continued)

```
# %% 2) visualize data

#  import plotting library
import matplotlib.pyplot as plt

# plot
plt.plot(
    data["concentration_g_l"],  # data on x-axis
    data["surface_tension_mN_m"]  # data on y-axis
    )
# show plot
plt.show()
```

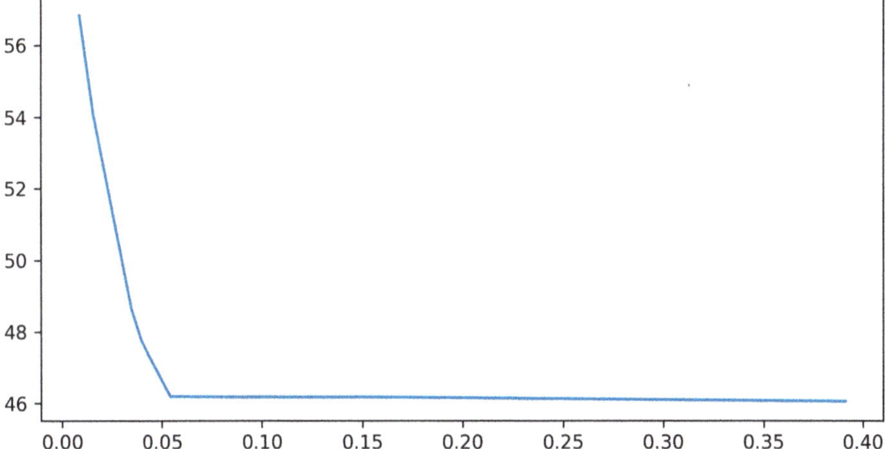

Figure 5.2: Exemplary surface tension isotherm plot, i. e., surface tension against concentration obtained from the basic `matplotlib.pyplot.plot`-function without specification of any other parameter. There are no axis labels or titles included using the default settings.

In order to learn the precise names of the `DataFrame`'s columns, we can use either the variable explorer or type `data.columns` in the console.

The result shown in Figure 5.2 is on the one hand not spectacular, but not bad on the other hand either for a few lines of code. In order to make it visually more appealing, we need to apply some further finishing touches:

- highlight the locations of data points via markers,
- remove the connection between the data points in order to highlight *measured* data (the lines are useful only as the famous *guide to the eye*),
- use logarithmic scaling on the concentration axis (this is the typical representation of surface tension isotherms in surface chemistry/material science),
- add axis labels,
- add a title for future reference, and
- optionally: add time-stamp information (keep in mind, that this information can also be drawn from the file creation date accessible via the files properties window).

Taking into account the suggested amendments, the code is slightly longer (see Listing 5.5), but the resulting Figure 5.3 is much closer to our visual expectations.

Listing 5.5: data_to_information.py (continued)

```
# specify column names to be used as x- and y-axes
x = "concentration_g_l"
y = "surface_tension_mN_m"

# specify label names
x_label = "Concentration / [g/l]"
y_label = "Surface tension / [mN/m]"

# plot "guide to the eye"
plt.plot(
    data[x],  # data on x-axis
    data[y],  # data on y-axis
    color="black",  # black color
    linestyle="-",  # solid line
    alpha=0.1  # opacity
    )

# plot data as points
plt.plot(
    data[x],  # data on x-axis
    data[y],  # data on y-axis
    marker="o",  # set marker symbol
    linestyle=" "  # no connecting lines
    )

# add - better formatted - axes labels
plt.xlabel(x_label, size=14)
```

```
plt.ylabel(y_label, size=14)

# logarithmic scaling on x-axis
plt.xscale("log")

# import module for handling dates
import datetime

# add timestamp of processing
plt.text(
    0.1,
    data[y].max()-1,
    f"Timestamp:\n{datetime.datetime.now().replace(microsecond=0)}"
    )

# add title
plt.title(file)
```

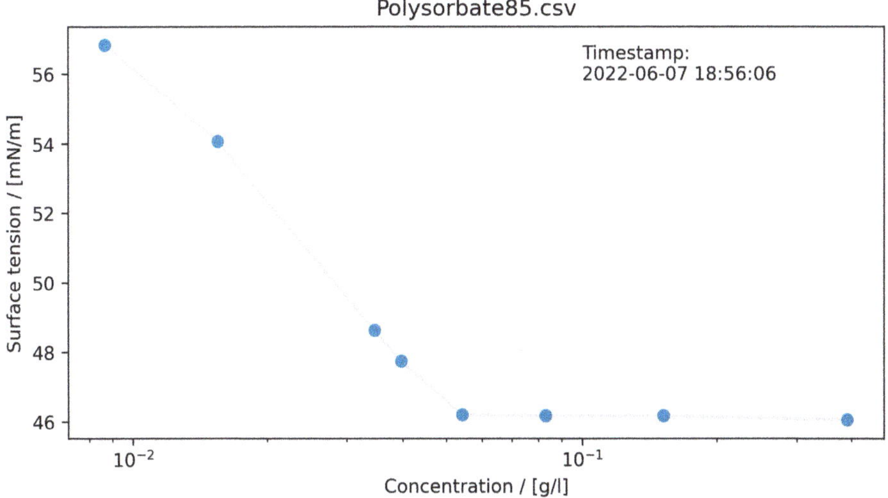

Figure 5.3: Exemplary surface tension isotherm plot obtained from `matplotlib` using some more customization. The x- and y-labels are named based on the respective column names plotted on the x- and y-axes. For context, the file name is set as title, and the timestamp of processing the experimental raw data file is added as text. To increase visual clarity, the experimentally measured surface tension-concentration pairs are represented via blue markers, connections between them are shown in a light grey colour as a *guide to the eye*.

After this first part of visual inspection of the results for this specific example, there are once again two possible options for the next steps:

- following the steps sketched in our pseudo-code in Listing 5.2 to go for *parameters* right away, or

- do the visualization for each and every of the files in the folder *raw_from_machine* and save the resulting figures to a new directory for visual inspection.

Again, this can be done according to your preferences. Getting a visual overview first might have some benefits if you expect some of your datasets corresponding to individual samples to differ strongly from others. This might have consequences on *how* we extract parameters, i. e., *information* from *data*.

Let me assure you that the examples used herein are sufficiently similar to each other to be analysed with the same algorithm. Nevertheless, I would like to show how the figure can be saved for visual inspection using the `savefig`-function in Listing 5.6.

Listing 5.6: data_to_information.py (continued)

```
# define output folder name
path_to_figure_export = os.getcwd() + os.sep + "_export_figures"
# create export path if it does not exist
if not os.path.exists(path_to_figure_export):
    # make folder
    os.mkdir(path_to_figure_export)

# save figure with base filename but other extension --> png
plt.savefig(
    path_to_figure_export + os.sep + file.replace(".csv", ".png"),
    bbox_inches="tight",  # remove whitespace around figure
    dpi=300  # set resolution
    )
```

5.5 Parameter extraction step by step

Now to the main part of this chapter: the extraction of *parameters* from *data*. As previously indicated, this relates to values that are known—in the respective field—to be relevant (or characteristic) for the description of a certain sample. Equally, you are free to extract parameters from the *data* that *you* consider relevant, e. g., for a specific application.

In our example, we will go for the extraction of the critical micellar concentration (CMC). This characteristic value is the concentration at which there is no further—significant—decrease in surface tension going from a lower concentrated to a higher concentrated surfactant solution, in other words, from left to right in the plot shown in Figure 5.3. Conceptually simple to understand, the question is how to "translate" that into some Python code.

What is your first impression looking at the Figure 5.3? Right, there are two regions with distinctly different slopes. Slopes can be obtained from the `gradient`-function of the `numpy` package. In almost any circumstance, the *alias* `np` is used for the `numpy` package. In order to get a better impression of the results obtained from this step, we just add the determined gradient to the plot to identify a criterion for distinguishing both ranges. Going from "left to right" on the concentration axis, the initially negative gradient calculated according to Listing 5.7 approaches a value more or less close to zero in the higher concentration range as shown in Figure 5.4.

Listing 5.7: data_to_information.py (continued)

```
# %% 3) get "state of the art" parameters

# module for calculating gradient (among others)
import numpy as np

# get gradient
gradient = np.gradient(
    data[y],  # y-axis data
    data[x]   # x-axis data
    )

# initialize figure
fig, ax = plt.subplots()

# introduce second axis
ax_gradient = ax.twinx()

# plot data as points on "ax"
ax.plot(data[x], data[y], "ko")
# add gradient on "ax_gradient" (secondary y-axis)
ax_gradient.plot(data[x], gradient, "rs-")

# logarithmic scaling on x-axis
plt.xscale("log")

# add - better formatted - axes labels
ax.set_xlabel(x_label, size=14)
ax.set_ylabel(y_label, size=14)
ax_gradient.set_ylabel(
                "Gradient of surface tension\n / [(mN/m)/(g/l)]",
                size=14
                )
```

For a clearer representation of both the original surface tension and concentration *data* and the calculated gradient in the same figure, we use a second y-axis obtained via the `twinx`-method of the `ax`-object.

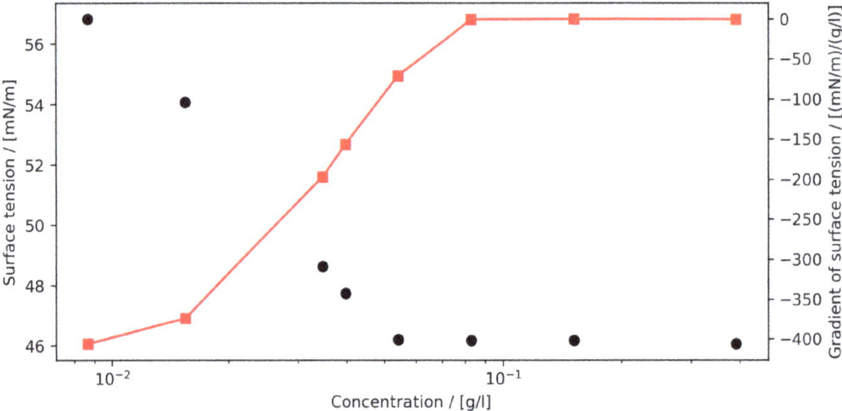

Figure 5.4: Exemplary surface tension isotherm (black circles) and corresponding gradient in surface tension plotted against concentration on a second y-axis (red squares) on the right- hand side of the plot.

For setting up a plot via `matplotlib`, typically the function `plt.subplots()` is used. It returns a tuple containing a figure and axes object(s). In many examples, the latter is “unpacked“ into the variables `fig` and `ax` using `fig, ax = plt.subplots()`. Having `fig` is useful for changing figure-level attributes such as size or for saving the figure as an image file later. Also, all `axes` objects (the objects that have plotting methods) have a parent `figure` object anyway.[2] The `twinx`-function is used to introduce a “second y-axis” within a plot. According to the documentation accessible by typing `ax.twinx?` in the console, it serves to create a new `axes` instance with an invisible x-axis and an independent y-axis positioned opposite to the original one.

With this plot in our hands, we have at least two options to identify the CMC-value of the respective surfactant. To be more precise, there are at least two options for identifying *a* CMC-value. As the number of measured points, i. e., surface tension–concentration pairs is not too large in our example, we are forced to choose one of the following assumptions:

- We actually did perform a measurement at *the* CMC-value. In this case, the task is as easy as picking the most appropriate value from the experimental data (see Figure 5.5)
  or
- We did not perform a measurement at *the* CMC-value. In this situation, the desired value is obtained from, e. g., the intersection point of two best-fit lines representing the mentioned high- and low-gradient concentration ranges. This obviously demands some more effort (see Figure 5.6).

2 https://stackoverflow.com/questions/34162443/why-do-many-examples-use-fig-ax-plt-subplots-in-matplotlib-pyplot-python

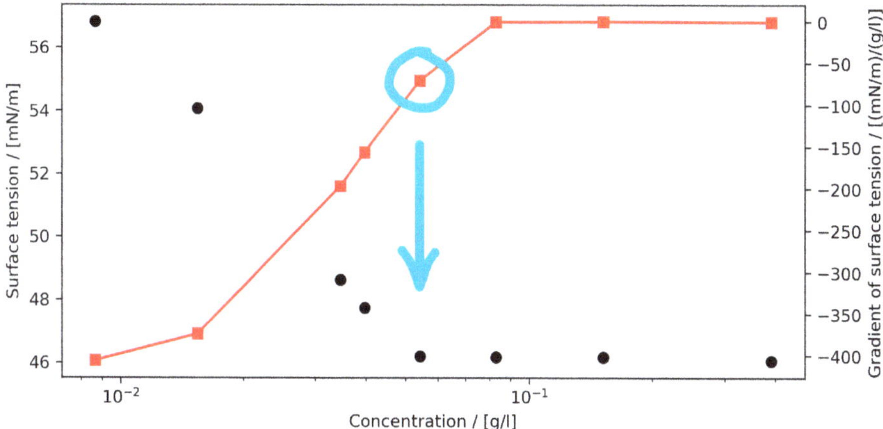

Figure 5.5: Option 1 for determining the CMC: It is obtained as the concentration at which a certain absolute value of the calculated gradient is exceeded. This threshold is to be set in the following and has to be tested on a reasonable number of experimentally gathered files for robustness.

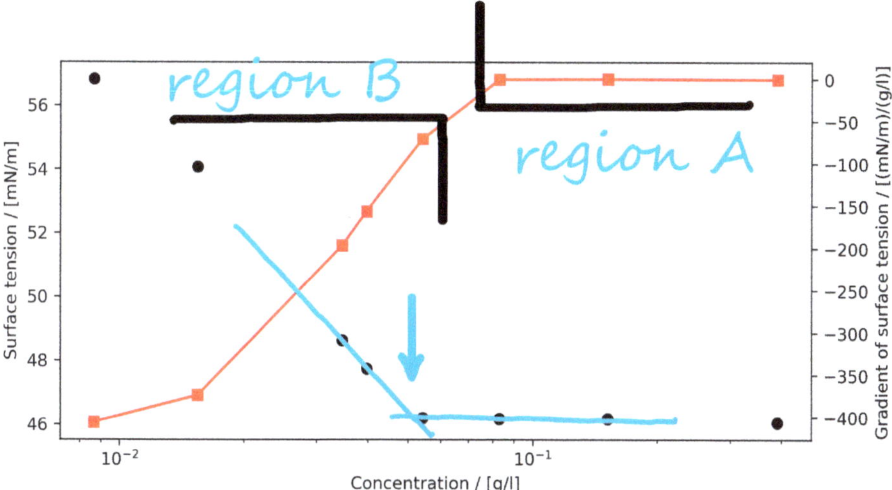

Figure 5.6: Option 2 for determining the CMC: The CMC is obtained as the intersection concentration between two best-fit lines corresponding to an above CMC region A and a below CMC region B. In this approach, the criterion for distinguishing between regions A and B via the slope can potentially be handled less strictly. Also, the obtained CMC will most likely not coincide with a measured surface tension–concentration value pair.

For clarity, the principles underlying the latter options to extract CMC-values from *data* are sketched in Figure 5.5 and Figure 5.6, respectively.

5.6 Selecting the CMC from experimentally determined values

Keeping things simple, we go for the first—easier—option as an initial guess. Should we realize from visual inspection, at a later stage, that this procedure leads to CMC-values too far off, we can rework then this part of the analysis. Using the selection approach, we first add the calculated `gradient` to our `data` as an additional column. Then, we apply a filter to it according to the best of our current knowledge. In our example, the calculated gradient values in the constant surface tension regime on the right-hand side of the plot have magnitudes below −1. The other points are associated with absolute gradient values around 100 and larger as visible from spyder's variable explorer or via a `print` statement. In order to arrive at a clear-cut separation, we might use a threshold value of, e. g., −50. As we can see from the conceptual Figure 5.5, we discard all surface tension–concentration pairs, i. e., rows having a gradient above this specified value. Next, we select the topmost row of the remaining `DataFrame` and finally select the concentration that we *use* as our CMC in the following. The process for selecting this value from an actual example is sketched in Figure 5.7.

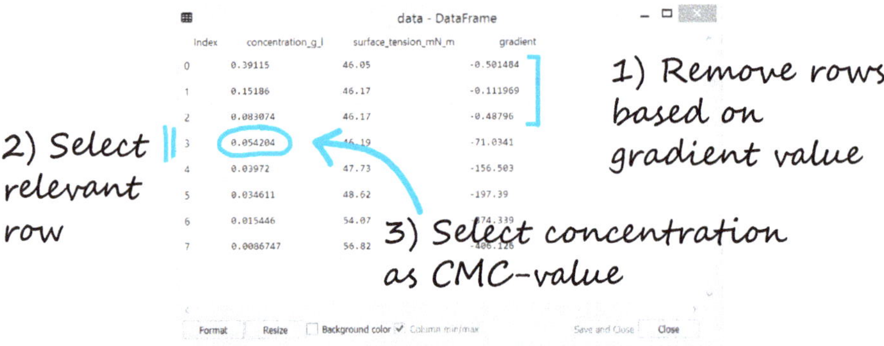
data - DataFrame

Index	concentration_g_l	surface_tension_mN_m	gradient
0	0.39115	46.05	-0.501484
1	0.15186	46.17	-0.111969
2	0.083074	46.17	-0.48796
3	0.054204	46.19	-71.0341
4	0.03972	47.73	-156.503
5	0.034611	48.62	-197.39
6	0.015446	54.07	[illegible]74.339
7	0.0086747	56.82	-406.126

Figure 5.7: Exemplary `DataFrame` with the columns for concentration, surface tension and gradient highlighted for determining the CMC according to the method sketched in Figure 5.5. To select the value of interest, we discard the first the rows having a low gradient value and conclude the analysis task by selecting the concentration value in the topmost row of the remainder.

In order to translate that in our script, we first use the `query`-method of our `data`. Therein, we specify that we only want to select rows in which the value in the gradient column is below −1. At this stage, the specific value is just an initial, and hopefully, appropriate guess for the purpose of identifying the CMC according to the previously introduced option 1. Next, we select the topmost row of the remaining data of interest, `doi`. We use the `head`-method of our `doi` with the argument 1 to obtain just the topmost row. The only remaining task is then to pick the numerical value in the column `concentration_g_l`. The corresponding code is shown in Listing 5.8.

Listing 5.8: data_to_information.py (continued)

```
# add "gradient" column to data
data["gradient"] = gradient

# drill down to "data of interest" as doi
# discard rows with gradient value below absolute threshold, i.e.
# selection of "below CMC" surface tension-concentration value pairs
doi = data.query("gradient < -1")

# select topmost remaing row
doi = doi.head(1)

# assume CMC as measured concentration from doi
cmc = float(doi[x])  # x = "concentration_g_l"

# show value in plot
ax.axvline(cmc, color="green", linestyle="--", label="cmc")

# add legend at location 2 ('upper left')
ax.legend(loc=2)

# save figure with base filename but other extension --> png
plt.savefig(
    path_to_figure_export + os.sep + file.replace(".csv", "_cmc.png"),
    bbox_inches="tight",
    dpi=300
    )

# show plot
plt.show()
```

Next, another visualization step might be a good choice. Does the concentration we extracted and will further use as CMC make sense? Is there some kind of flaw in our algorithm so far? Highlighting this characteristic concentration via a vertical line is both easy and will give some instant feedback. Taking a look at Figure 5.8, we immediately see that we are approximately right, at least for this example.

Again, saving this plot as a figure in the previously mentioned directory might be helpful to obtain a visual impression of the validity of this parameter-extraction step. Of course, saving images at this point is not necessarily required. You can also perform this visual quality control only after storing these characteristic values to a certain folder structure, file or database and carry out this step then. This, however, might lead to some additional work since it requires removing the erroneously stored characteristic value from the storage system, redoing the evaluation and carrying out the saving step once more. Another reason to do this check early in the process is to ensure the quality of your parameters on the *information* level right from the start. You do not want to worry about the correctness of your extracted parameters once

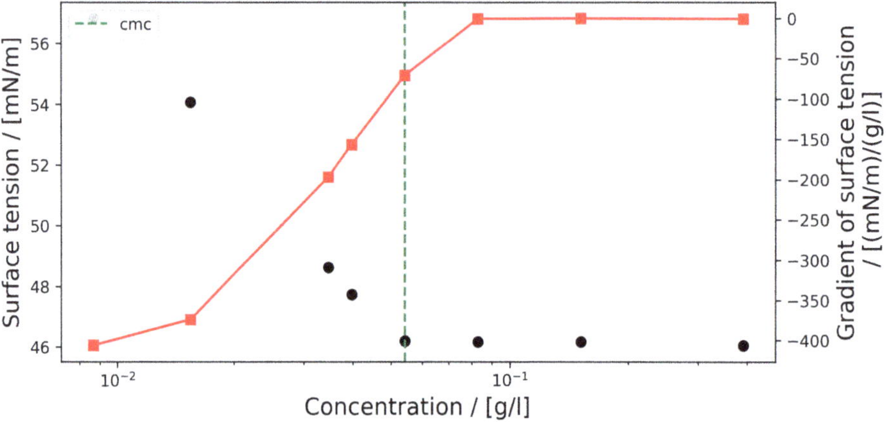

Figure 5.8: Exemplary surface tension isotherm (black circles) and corresponding gradient (red squares) on a second y-axis. Furthermore, the CMC-value determined according to the previously described procedure is indicated via a vertical dashed line.

you are on the level of actually using them, i. e., making the step from *information* to *knowledge* and beyond.

5.7 Building the `data_to_information` module

Just as in the previous chapter, we now want to shorten our code, thereby making it simpler and clearer. Therefore, we define another module in our `surface_tension` package. According to the current step, we will name the module `data_to_information` (see Listing 5.9) and declare the custom functions `plot` and `get_cmc` within.

Listing 5.9: surface_tension/data_to_information.py

```
import matplotlib.pyplot as plt
import numpy as np
import os


#
# plot surface tension isotherm data
#
def plot(data, sample, cmc=None, export_path=None, show_info=True):
    """
    plot surface tension isotherm data and optionally CMC value

    Parameters
    ----------
```

```
    data : pd.DataFrame
        surface tension as obtained from file_to_data's function
        get_data_from_experimental_string.
    cmc : float | int, optional
        CMC value to be highlighted as vertical line.
        The default is None.
    show_info : bool, optional
        flag for showing intermediate processing results.
        The default is True.

    Returns
    -------
    None.

    """

    # specify column names to be plotted on x- and y-axes
    x = "concentration_g_l"
    y = "surface_tension_mN_m"

    # plot "guide to the eye"
    plt.plot(
        data[x],  # data on x-axis
        data[y],  # data on y-axis
        "k-",  # solid (-) line without marker in black (k) color
        alpha=0.1  # opacity
        )

    # plot data as points
    plt.plot(
        data[x],  # data on x-axis
        data[y],  # data on y-axis
        marker="o",  # set marker symbol
        linestyle=" "  # no connecting lines
        )

    # optional plot of CMC as vertical line
    if cmc:
        # vertical line
        plt.axvline(cmc, color="red")

    # add - better formatted - axes labels
    plt.xlabel(x.replace("_", " "))
    plt.ylabel(y.replace("_", " "))

    # logarithmic scaling on x-axis
    plt.xscale("log")

    # add legend
```

```
    plt.legend()

    # add title
    plt.title(sample)

    # save image, if path specified
    if export_path:
        # save figure with base filename but other extension --> png
        plt.savefig(
            export_path + os.sep + sample + ".png",
            bbox_inches="tight"  # get rid of whitespace
            )

    # show plot
    plt.show()

    # return
    return


#
# extract CMC value
#
def get_cmc(data, show_info=True):
    """
    extract CMC value from DataFrame

    Parameters
    ----------
    data : pd.DataFrame
        surface tension as obtained from file_to_data's function
        get_data_from_experimental_string.
    show_info : bool, optional
        flag for showing intermediate processing results.
        The default is True.

    Returns
    -------
    CMC as float.

    """

    # specify relevant column names for the analytical task
    x = "concentration_g_l"
    y = "surface_tension_mN_m"

    # get gradient
    gradient = np.gradient(
        data[y],  # x-axis data
```

```
        data[x]  # y-axis data
        )

    # add "gradient" column to pd.DataFrame "data"
    data["gradient"] = gradient

    # info
    if show_info:
        print(data)

    # drill down to "data of interest" as doi
    # discard rows with gradient value above threshold, i.e.
    # restrict to rows with gradient below threshold
    doi = data.query("gradient < -1")

    # select topmost remaing row
    doi = doi.head(1)

    # assume CMC as measured concentration from doi
    cmc = float(doi[x])
    # info
    if show_info:
        print("-->", cmc)

    # return CMC-value
    return cmc
```

5.8 Using the custom data_to_information module

With this module in hand, we can dramatically shorten the code for extracting the characteristic CMC-value from our experimental raw data as shown in Listing 5.10.

Listing 5.10: data_to_information_via_function.py

```
# import for folder management
import os

# get current directory
current_dir = os.getcwd()
# derive path to data starting from current directory of the script
path_to_data = current_dir + os.sep + "raw_from_machine"

# define output folder name
path_to_figure_export = os.getcwd() + os.sep + "_export_figures"
# make path if it does not exist
if not os.path.exists(path_to_figure_export):
    # make folder/directory
```

```
    os.mkdir(path_to_figure_export)

# loop path to data to show list of files
for file in os.listdir(path_to_data):
    # show filename
    print(file)

    # use filename without extension as sample name
    sample=file.split(".csv")[0]

    # get full filename including path
    file_path = path_to_data + os.sep + file


# %% 1) Read Experimental File content, i.e. "data"

# import custom modules "file_to_data" and "data_to_information"
# modules from "surface_tension" package
from surface_tension import file_to_data
from surface_tension import data_to_information

# get file content as string
file_content_string = file_to_data.read_file_content_as_string(
                                                            file_path
                                                            )

# extract "data" from file content
data = file_to_data.get_data_from_experimental_string(
            file_content_string,
            show_info=False
            )

# get CMC
cmc = data_to_information.get_cmc(data, show_info=False)

# plot
data_to_information.plot(
        data,
        sample, # sample name
        cmc=cmc, # add CMC information
        export_path=path_to_figure_export # target directory
        )

# remove no longer required variables
del file_path, file_content_string
```

5.9 Wrap up

This chapter introduced a suggestion for a folder structure to hold the experimental raw data files. For laying out the required processing steps, pseudo-code demonstrating the major steps was shown. Building on the already defined `file_to_data` module, the *data* was accessed, followed by a basic visualization to identify options for the extraction of *parameters*, i. e., values characteristic of the sample on the *information* level. This procedure was shown step by step based on the *selection* of an experimentally characterized concentration such as the CMC of the corresponding aqueous surfactant solution.

The chapter concludes with the creation and exemplary use of the freshly introduced module `data_to_information` based on the functions developed herein.

6 Where to put data and information

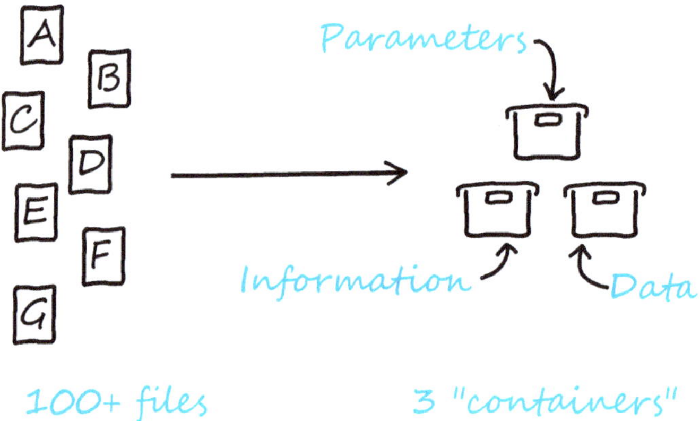

Now that we are able to readily access both *data* and *information* using dedicated scripts as described in the recent chapters, we face the question where to actually put them. In other words: What do we do with the obtained *data* and *information*? Where should we store and keep it for later use? The suggested solutions fall into three distinct categories:

- Relying on a properly sorted folder structure populated based on scripts applying strict naming conventions. The files should not be opened and modified manually. The drawback of this very simple approach is that it is not readily scalable to vast amounts of *data* and *information*. For a number of samples in the range of tens or hundreds, however, this should be sufficient. In the latter situation, the more critical aspect is expected to be the—probably missing—discipline in file handling: *Should* not be modified manually here reads *must* not be modified manually—except when you are certain that you know what you are doing.
- Intermediate solutions rely on a local database such as SQLite or Microsoft® Access®. These approaches make modification of *data* or *information* by mistake much less likely. Using SQLite has the further benefit of seamless interaction with Python. Therefore, this approach will be shown in greater detail later in this chapter.
- In a larger context, hosted relational database management systems, such as Microsoft SQL Server or PostgreSQL (free and open-source relational database management systems), can be used to store larger amounts of data accessible for multiple users.

https://doi.org/10.1515/9783110788433-006

Structured Query Language (SQL) is a standard language for accessing and manipulating databases.[1] It is designed for managing data held in a relational database management system, or for stream processing in a relational data-stream management system. It is particularly useful in handling structured data.[2]

SQLite is an in-process library that implements a self-contained, serverless, zero-configuration, transactional Structured Query Language database engine. The code for SQLite is in the public domain and is thus free to use for any purpose, commercial or private. Unlike most other SQL databases, SQLite does not have a separate server process. SQLite reads and writes directly to ordinary disk files. A complete SQL database with multiple tables, indices, triggers, and views, is contained in a single disk file. The database file format is cross-platform. These features make SQLite a popular choice as a database engine.[3]

Independent of which approach you want to use for your project, the general idea is merely to *provide some space* for your *data* and *information* in order to have it readily available for your downstream processes. In the following, saving your results to an organized folder structure via a script and saving them to a SQLite database via the `SQLAlchemy`-package will be presented.

6.1 The minimal option: organized folder structure

Using an organized folder structure is probably the most basic way for storing your raw data, *data*, *information* and results. Nevertheless, it might be an absolutely valid option for your—probably smaller—project. If you are aiming to collect a lot of data for tens of thousands of samples, this is probably not the best way to go. If you are, on the other hand, expecting to perform measurements for only some tens or even hundreds of samples, it might be worth a try. Keep in mind that this way requires strict discipline concerning touching, i. e., manually modifying the files. If you cannot make sure that you—and also colleagues contributing to the project—cannot refrain from opening, inspecting and possibly even editing the files (in an improper) manner, this method for storing is prone to error. Barring that scenario, you will probably be fine.

In order to keep things nice and clean, the following procedure building on top of the previously defined modules of our `surface_tension` package are suggested. Once more, some pseudo-code is used to pin down the key steps as shown in Listing 6.1.

---

1 https://www.w3schools.com/sql/sql_intro.asp

2 https://en.wikipedia.org/wiki/SQL

3 https://www.SQLite.org/about.html

Listing 6.1: pseudo_code_where_to_put_data_and_information_I.py

```
#
# Loop files in raw data folder

    #
    # 1) Read

    #
    # 2) Get and save data

    #
    # 3) Get and save information

    #
    # 4) Optional: Save plots for visual inspection (quality control)

    #
    # 5) Move file from raw data folder to raw data archive
```

In contrast to the previously shown code snippets, we will now apply all of the so-far defined processing steps, i. e., reading, extraction of both *data* and *information* and plotting to each raw data file contained in the directory *raw_from_machine*. Furthermore, we will move the processed file to a newly defined archive folder, *raw_archive*. This file transfer brings along both an advantage and a potential risk.

- Advantage: There will be no duplicates or rather multiple analyses of one and the same raw data file running the analysis script again at a later point in time. This is *the* typical scenario of ongoing lab work, where you expect, e. g., a fixed number of files to be moved to the raw data folder *raw_from_machine* each day.
- Risk: There will be no duplicates or rather multiple analyses of one and the same raw data file running the analysis script again at a later point in time. The very same behaviour might as well be an issue if you modify one of the functions in the modules `file_to_data` or `data_to_information`. Furthermore, your file management system will now allow you to place a file with the very same name you have in your archive to the raw data folder (you can consider this directory as some kind of *inbox*). Running the script again will overwrite your existing file in the archive—unless you check for this source of error in the Python code and issue a warning or feedback message to the user.

Setting the potential difficulties aside, the key components of an organized folder structure are sketched in Figure 6.1.

As a first step, we generate the folder structure according to Figure 6.1. As seen from Listing 6.2, we first check if the respective directory already exists via the `os.path.exists`-function and generate the non-existent folders via `os.mkdir` if appli-

script.py
_data
A.csv
B.csv
_information
A.csv
B.csv
_parameters
A.csv
B.csv

Figure 6.1: Schema of the proposed organized folder structures for splitting experimental raw data files into *data* and *information* using the previously built `surface_tension` package. Accordingly, there are folders containing *data*, *information* and *parameters* on the same level as the Python-script providing content to them based on the experimental raw data folder's content. Each of the folders contains files with the same name to ensure easy mapping.

cable. This part of the script will of course only be run during the *very first* execution of the code.

Listing 6.2: where_to_put_data_and_information_I.py

```
# import for folder management
import os
# import for data handling
import pandas as pd
# import for moving files and other file related tasks
import shutil
# import custom modules from "surface_tension" package
from surface_tension import file_to_data
from surface_tension import data_to_information


# get current directory
current_dir = os.getcwd()
# derive path to data starting from current directory of the script
path_to_data = current_dir + os.sep + "raw_from_machine"

# define output folder names
path_to_figure_export = os.getcwd() + os.sep + "_export_figures"
path_to_data_export = os.getcwd() + os.sep + "_data"
path_to_information_export = os.getcwd() + os.sep + "_information"
path_to_parameters_export = os.getcwd() + os.sep + "_parameters"
path_to_raw_archive = os.getcwd() + os.sep + "raw_archive"

```

```
# for each output/export path: make path if it does not exist
for _p in [path_to_figure_export, path_to_data_export,\
           path_to_information_export, path_to_raw_archive,\
           path_to_parameters_export]:
    # check if path exists
    if not os.path.exists(_p):
        # make folder
        os.mkdir(_p)
```

Next, we will go for the looping part, i. e., go over all the files in our raw data folder *raw_from_machine* (see Listing 6.3). In particular, we will loop over the `list` returned from the `os.listdir`-function. Here, the return value of this function is not stored in a variable but used directly in the `for`-loop.

! Add a `break` statement at the end of a `for` loop while building your loop processing to stop immediately after the first iteration. When satisfied with the processing of the first element, remove the `break` and check the overall results.

Listing 6.3: where_to_put_data_and_information_I.py (continued)

```
for file in os.listdir(path_to_data):
    # info
    print("Processing file", file)

    #
    # 1) Read
    _raw_string = file_to_data.read_file_content_as_string(
        path_to_data + os.sep + file
        )

    #
    # 2) Get and save data
    # get data
    data = file_to_data.get_data_from_experimental_string(
        _raw_string,
        show_info=False
        )
    # save data
    data.to_csv(
        path_to_data_export + os.sep + file,
        index=False  # do not export index
        )

    #
    # 3) Get and save information
    # get information
    information = file_to_data.\  # multi-line function via "\"
                 get_information_from_experimental_string(
                     _raw_string,
```

```
                    show_info=False
                )
    # add further "known" parameters not captured in the results file
    information = file_to_data.add_further_information(
                information,
                medium="water",  # as solvent/medium information
                temperature=23  # use "room temperature" as best guess
                )
    # save information
    information.to_csv(
        path_to_information_export + os.sep + file,
        index=False  # do not export index
        )

    # get CMC as a parameter (also on information level)
    cmc = data_to_information.get_cmc(
        data,
        show_info=False
        )
    # add CMC to "parameters" DataFrame and save it
    parameters = pd.DataFrame({"cmc_g_l" : cmc}, index=[0])
    parameters.to_csv(
        path_to_parameters_export + os.sep + file,
        index=False  # do not export index
        )

    #
    # 4) Optional: Save plots for visual inspection (quality control)
    data_to_information.plot(
        data,
        file.replace(".csv", ""),  # clean file name without extension
        cmc=cmc,
        export_path=path_to_figure_export
        )

    #
    # 5) Move file from raw data folder "raw_from_machine" to
    # raw data archive folder "raw_archive"
    # define filename in archive folder
    file_in_archive = path_to_raw_archive + os.sep + file
    # move only if no file of the same name in archive
    if not os.path.exists(file_in_archive):
        # move raw file there
        shutil.move(
            path_to_data + os.sep + file,
            file_in_archive
            )
    else:
        # info
        print("File", file, "alredy exists in", path_to_raw_archive)
```

6.2 Adapting module functions

Let's take a moment to inspect the figures exported to the folder *_export_figures*. From Figure 6.2 it is quite obvious that some minor adaptions are still required. For two of the samples considered so far, the determined critical micellar concentration (CMC) as indicated by the vertical line differs more from the visually observable "kink" than we would like it to.

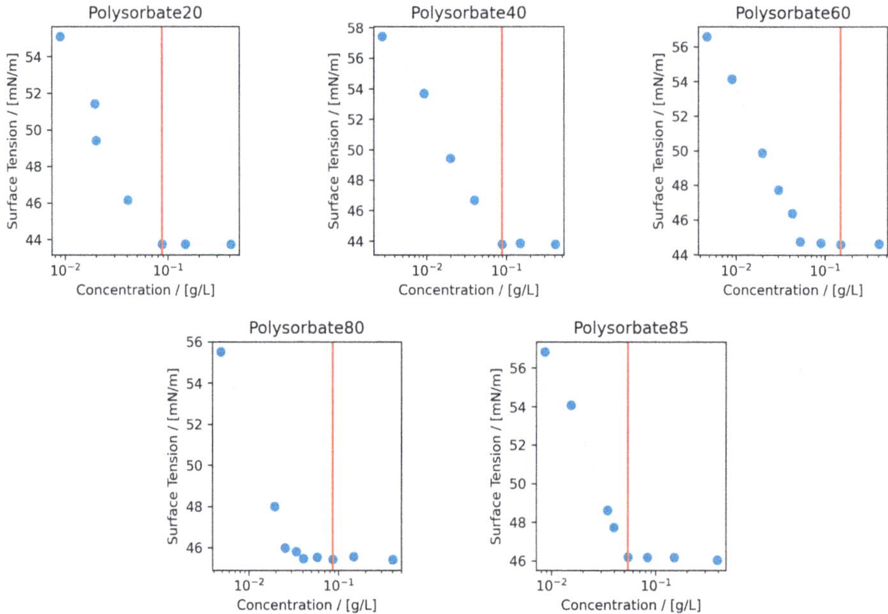

Figure 6.2: Visual inspection of the CMC-values obtained from the previously described procedure. For two samples (Polysorbate60 and Polysorbate80), we do not see quite the expected result. This means that some parameter fine-tuning is required. The indicated CMC values were determined using a gradient threshold value of –1.

In order to compensate for this undesired behaviour, we have to go back to our analysis steps related to the CMC. As indicated in Figure 5.8, the only tunable value we use in our analysis is the threshold or cut-off value of the gradient above which surface tension–concentration pairs are discarded. The goal of the following adaptions is to introduce an argument in the `get_cmc`-function of the `data_to_information` module that allows for setting this gradient threshold value upon calling the function. Accordingly, we need to set this value tunable in our `get_cmc`-function. Therefore, we introduce the optional parameter `gradient_threshold` with a default value of –1, i. e., the value used so far to the `get_cmc` function of the `data_to_information` module. Only two lines of code have to be adjusted in to make this change come to life in the data_to_information module, as shown in Listing 6.4. We introduce the optional keyword argument gradient_threshold in the function definition and filter for this value within the query-function in line 131 of the displayed code snippet. The remainder of the function is not touched.

Keep in mind to also consider the additional keyword gradient_threshold in the documentation of the function accordingly.

Listing 6.4: surface_tension/data_to_information.py

```
92  def get_cmc(data, gradient_threshold=-1, show_info=True):
93      """
94      extract CMC value from DataFrame
95  
96      Parameters
97      ----------
98      data : pd.DataFramne
99          data abtained from
100         file_to_data.get_data_from_experimental_string.
101     gradient_threshold: int, float
102         gradient threshold above which values are discarded
103     show_info : bool, optional
104         Flag for showing information. The default is True.
105 
106     Returns
107     -------
108     CMC as float.
109 
110     """
111 
112     # specify column names to be used as x- and y-axes
113     x = "concentration_g_l"
114     y = "surface_tension_mN_m"
115 
116     # get gradient
117     gradient = np.gradient(
118         data[y],  # x-axis data
119         data[x]  # y-axis data
120         )
121 
122     # add "gradient" column to data
123     data["gradient"] = gradient
124 
125     # info
126     if show_info:
127         print(data)
128 
```

```
    # drill down to "data of interest" as doi
    # discard rows with gradient value below threshold
    doi = data.query("gradient < @gradient_threshold")

    # select topmost remaing row
    doi = doi.head(1)

    # assume CMC as measured concentration from doi
    cmc = float(doi[x])
    # info
    if show_info:
        print("-->", cmc)

    # return CMC-value
    return cmc
```

Also, our overall processing script requires only a change in one line as shown in Listing 6.5: introducing the `gradient_threshold`-keyword with a value other than the default of −1.

Listing 6.5: where_to_put_data_and_information_I_flexible_threshold.py

```
    # get CMC as a parameter (also on information level)
    cmc = data_to_information.get_cmc(
        data,
        gradient_threshold=-20,  # set threshold here
        show_info=False
        )
```

Tuning the newly introduced parameter `gradient_threshold`, we end up with the results of Figure 6.3. Apparently, the numerically determined CMC now agrees much more with the position of the kink in the surface tension versus concentration plot.

6.3 Just one data point...

Of course, one might argue that the determined CMC at the bottom left of Figure 6.3 (plot titled Polysorbate80) is not where it "should" be according to our scientific understanding. However, there are some—and maybe even good—reasons to leave the extracted CMCs as they are right now:

– The algorithm for extracting the CMC according to the gradient works just as intended. Furthermore, it is not *clearly wrong*, and possible deviations to values expected from visual observation might originate from numerically looking at

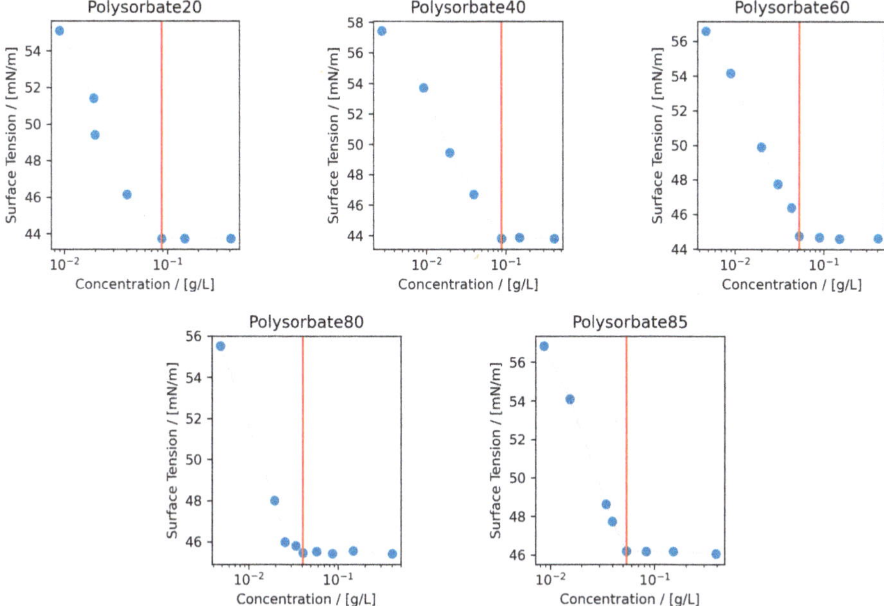

Figure 6.3: Visual inspection of the CMC-values obtained from the previously described procedure after some parameter fine-tuning. There is better agreement between the “visually expected” CMC and the value obtained from the `get_cmc`-function of the modified `data_to_information` module. The indicated CMC values were determined using a gradient threshold value of −20.

the data via another approach (such as the intersection option described in Figure 5.6). Also, this more complex approach will be subject to other sources of error.

- Another method of data collection might lead to a CMC closer to the expectation. To be more precise, using a tighter “sampling frequency”, i. e., more surface tension–concentration pairs leads to a more continuous course of the plots, thereby granting higher validity to the applied gradient and CMC *selection* approach.
- We need to keep in mind that also our data is subject to experimental error. The fewer data points, the more prominent is the influence of just and individual error. Assuming that the data point highlighted via a square marker in Figure 6.4 is either too low or too far to the left (due to an erroneous measurement or sample preparation), we find that the determined CMC is probably not so bad.

In order to see how much of a difference just *one* data point makes in the visual perception of the apparent correctness of the extracted CMC, see Figure 6.4.

With this in mind, we can either go back to the lab and reissue the measurement or just accept the extracted value of the CMC as it is. In the present case, we will go with the latter option.

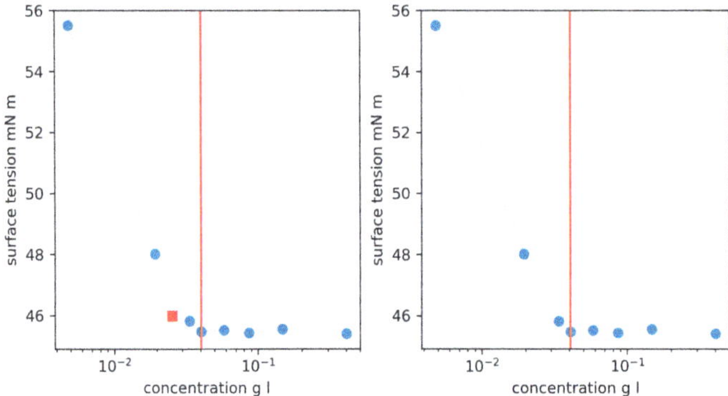

Figure 6.4: The difference just one data point makes. With the additional value (red square), the obtained CMC value indicated by the vertical line appears to be quite off. Without this point, the determined CMC seems to be a good fit.

6.4 Collecting *data* and *information* from multiple files

Zooming out from the individual raw data file-processing perspective, where are we right now? Starting from our experimental raw data, we so far ended up with equally named files in the folders *_data*, *_information* and *_parameters*. To make this *spilled* data and information available for convenient plotting via typically available programs, such as Microsoft® Excel®, Origin,[4] tableau[5] or the like, we need to combine their respective contents. You guessed correctly: It's once more time for looping over those files and merging the folder contents into individual files.

Collecting the contents of the folder *_data* according to the principle outlined in Figure 6.5 is shown in Listing 6.6.

e.g. for "_data" folder

_data
A.csv
B.csv
C.csv
"collect"
_DATA.csv

Figure 6.5: Purpose of the *collect_data* Python script. Contents of the individual *csv*-files in the data folder are collected in a summary file *_DATA.csv*.

4 https://www.originlab.com/

5 https://www.tableau.com/

Listing 6.6: collect_data.py

```
# import for folder management
import os
# for data handling
import pandas as pd

# get current directory
current_dir = os.getcwd()

# specify folder to be looped
path_to_data = os.getcwd() + os.sep + "_data"

# loop folder
for _file in os.listdir(path_to_data):
    # info
    print(_file)
    # read file contents as pd.DataFrame
    _this_data = pd.read_csv(
        path_to_data + os.sep + _file
        )
    # add sample or ID information
    _this_data["id"] = _file

    # build overall DataFrame
    try:
        data = data.append(_this_data)
    except:
        data = _this_data

# save collected results to csv
data.to_csv(
    current_dir + os.sep + "_DATA.csv",
    index=False  # do not export index
    )
```

As a result, we end up with a plain *csv*-file that can easily be imported into almost any other program for further processing. The first lines of the thereby obtained file *_DATA.csv* are printed in the following:

```
concentration_g_l,surface_tension_mN_m,id
0.40949,43.73,Polysorbate20.csv
0.14797000000000002,43.73,Polysorbate20.csv
0.087851,43.73,Polysorbate20.csv
0.040554,46.13,Polysorbate20.csv
0.019972,49.4,Polysorbate20.csv
0.019441,51.4,Polysorbate20.csv
```

```
0.0089791,55.07,Polysorbate20.csv
0.4079,43.78,Polysorbate40.csv
0.14911,43.84,Polysorbate40.csv
0.08899299999999999,43.78,Polysorbate40.csv
0.03972,46.68,Polysorbate40.csv
0.019793,49.43,Polysorbate40.csv
0.0092199,53.69,Polysorbate40.csv
0.0027936,57.4,Polysorbate40.csv
0.40602,44.6,Polysorbate60.csv
0.14911,44.58,Polysorbate60.csv
0.08899299999999999,44.67,Polysorbate60.csv
0.052587,44.74,Polysorbate60.csv
0.043003,46.37,Polysorbate60.csv
```

The file contains three columns: concentration in $g\,L^{-1}$, surface tension in $mN\,m^{-1}$ and the sample name as specified via the file name. In short: pretty much everything you will need for comparing your *data* according to a specific request or an idea coming to your mind.

! Relying on the file/folder structure provided by your machine's operating system, *unique* naming of your samples is ensured. Saving two result files having the same name in a given directory, e. g., *raw_from_machine*, is simply not possible. This comes with the benefit of having an *unique identifier*.

Accordingly, files collecting *information* and *parameters* can be generated. Therefore, just one additional layer of looping is required in the previous script. The corresponding code is shown in Listing 6.7. Therein, the contents of the individual files contained in the folders *_data*, *_information* and *_parameters* are read, combined and saved to the respective files *_DATA.csv*, *_INFORMATION.csv* and *_PARAMETERS.csv*. For this purpose a double for-loop structure is applied:

- an "outer" loop iterating the folders, and
- an "inner" loop iterating the individual files contained within.

Listing 6.7: collect_from_all_folders.py

```
# import for folder management
import os
# for data handling
import pandas as pd

# get current directory
current_dir = os.getcwd()

# specify folders to be looped
```

```
path_to_data = os.getcwd() + os.sep + "_data"
path_to_information = os.getcwd() + os.sep + "_information"
path_to_parameters = os.getcwd() + os.sep + "_parameters"

# loop #1 | "outer" loop
# "zip" together folder and file name
for _folder, _target in zip(
        [path_to_data, path_to_information, path_to_parameters],
        ["_DATA.csv", "_INFORMATION.csv", "_PARAMETERS.csv"]
        ):
    # info
    print("Merging contents of", _folder, "to", _target)

    # loop #2 | "inner" loop
    # loop current folder
    for _file in os.listdir(_folder):
        # info
        print("   -", _file)
        # read file contents as pd.DataFrame
        _this_data = pd.read_csv(
            _folder + os.sep + _file
            )
        # add sample or ID information
        _this_data["id"] = _file

        # build overall DataFrame
        try:
            data = data.append(_this_data)
        except Exception as e:
            # info
            print(e)
            data = _this_data

    # save collected results to csv
    data.to_csv(
        current_dir + os.sep + _target,
        index=False  # do not export index
        )

    # info
    print("--> SAVED!\n")

    # remove "data" variable
    del data
```

To take this one step further, we can also run this *collection* step facilitated by *collect_from_all_folders.py* at the end of the processing routine outlined in Listing 6.5. Other than what has been shown so far, I would like to introduce an alternative to running the code as a function: We run the script using the os.system-function as shown in Listing 6.8.

Listing 6.8: run_collection_script.py

```
# import
import os

# run "collecting script" using Python
os.system("python collect_from_all_folders.py")

# info
print("--> Done")
```

The os.system-function executes a command in a subshell. Here, we want Python to run the script *collect_from_all_folders.py*. As stated before, you could as well define a function serving the purpose of *collect_from_all_folders.py*. Running the ready-made file is just another alternative to be considered in your toolbox.

6.5 More advanced: using SQLite

Aside the minimalistic *organized folder structure* approach from the previous section, there is another possibility coming naturally with Python: using an *SQLite*-database. It does not require installation of servers and is based on local files. Furthermore, interfacing between Python and SQLite is done via SQL.

The basic structure of a database is the following: Within one database, there are typically multiple tables. Each of the tables has a previously defined structure, i. e., the variable types therein are to be defined *before* the database is actually filled. Working with a database typically comprises the following steps:[6]

- create a database,
- create one or more tables,
- insert entries,
- optional: modify entries and
- select entries for inspection.

Within this book, we will rely on the SQLAlchemy package for the manipulation of SQLite-databases: a toolkit giving the full power and flexibility of SQL to Python developers.[7] It provides a full suite of well-known enterprise-level persistence patterns,

6 https://www.informatik-aktuell.de/betrieb/datenbanken/datenbanken-mit-python-und-SQLite.html

7 https://www.sqlalchemy.org/

designed for efficient and high-performing database access, adapted for a simple language [1].

Reading further, SQLAlchemy will make you stumble over the concept of *database abstraction* quite soon. In other words, this means a system for database communication that conceals the majority of details of how data is stored and queried. SQLAlchemy takes the position that the developer must be willing to consider the *relational* form of the data.[8]

In short: Independently of whether you are using SQLite or any other hosted SQL-database system, interfacing with the latter will feel quite the same through the use of SQLAlchemy.

6.6 Database creation

So let's get started with creating the database in the first place. To begin with, we have to define an Engine, which is responsible for managing the connection to the database. The Engine achieves this by incorporating a database connection pool and a database specific Dialect layer to translate the SQL expression language into database-specific SQL. Another class to become familiar with is MetaData. It collects objects that describe among others, the tables and indexes of the database. To establish a connection between the Engine and the MetaData, we will need to *bind* them. Setting up an empty SQLite-database is as easy as shown in Listing 6.9.

Listing 6.9: create_sqlite.py

```
# module for handling SQLite
import sqlalchemy

# create connection to the on-disk database "surface_tension.db"
engine = sqlalchemy.create_engine(
            'sqlite:///surface_tension.db',
            echo=True
            )

# create bound metadata
metadata = sqlalchemy.MetaData(engine)
```

6.7 Create tables

Next, we will need to define the tables contained in our database. From the previous section, we already know which contents they should hold. In short, our task in the

8 See http://aosabook.org/en/sqlalchemy.html

present step is to transfer the contents of our files *_DATA.csv*, *_INFORMATION.csv* and *_PARAMETERS.csv* to the respective tables within our `SQLite`-database.

! Transferring the contents of the previously mentioned *csv*-files does not necessarily mean that these files have to exist. Rather than collecting or rereading these files, the relevant *data*, *information* and *parameters*, could be inserted into the database right away in a script similar to Listing 6.7. With this approach, you would get rid of the intermediate "*.csv-step*". For reasons of clarity, this step is described more elaborately in the present discussion.

As indicated in Listing 6.10, we create a first table named `table_parameters` having two columns: the sample identification number (here we use the sample's file name for simplicity) and the extracted CMC. Further, we specify the sample name as a *primary key*.

---

i A *primary key* is the (one or multiple) column(s) that contains values that uniquely identify each row in a table. Primary keys necessarily have to contain unique values and cannot be null. A primary key can also be constructed from multiple columns.

Here, we use the sample name as the primary key. In our basic example, the name of the surfactant alone is enough to make it unique in the table. If you intend to investigate multiple batches of one type of surfactant, the primary key would be given by the combination of surfactant name and batch number. Having both surfactant type and batch specified, the measurement is defined uniquely. If multiple replicates of one and the same surfactant batch are of interest, the primary key would have to be extended by, e. g., a running number starting from 1. Each sample can only have one CMC under the conditions of the conducted experiments. Therefore, the table will contain as many rows as samples.

Next, we create the tables holding both our *data* and *information*. According to the structure in the *csv*-file, the `data` table will have the three columns:

- sample name,
- concentration and
- surface tension.

For the table `table_information`, we will use a modified version of the file *_INFORMATION.csv*'s column structure. Instead of using the original columns *parameter*, *value* and *id*, we will construct a reformatted table with the sample names as rows, the entries of the *parameter* column as new column names and the respective values from the *value* column. In short, the contents of *_INFORMATION.csv* are pivoted (see example in Subsection 2.3.5.1 for details). A conceptual sketch is given in Figure 6.6.

Furthermore, we declare the *id* column of `table_information`, i. e., our sample names, as a *foreign key* of the table `table_parameters`. The hierarchy between the tables is shown in Figure 6.7.

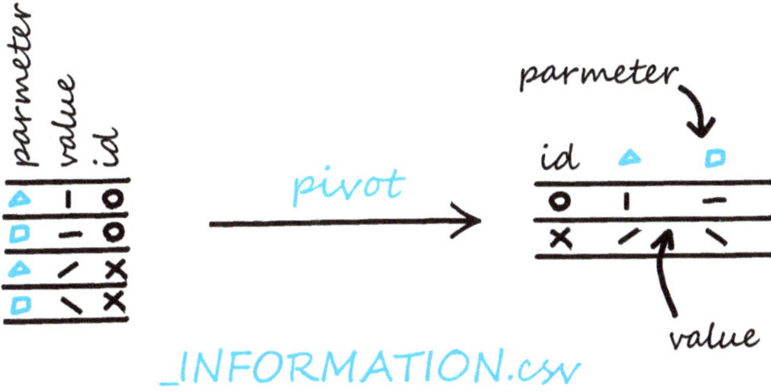

Figure 6.6: Schematic pivoting of the *_INFORMATION.csv* file's contents carried out in the following. Therein, the entries of the column *parameter* are used as column headers. Values populating the new table are taken from the *value* column.

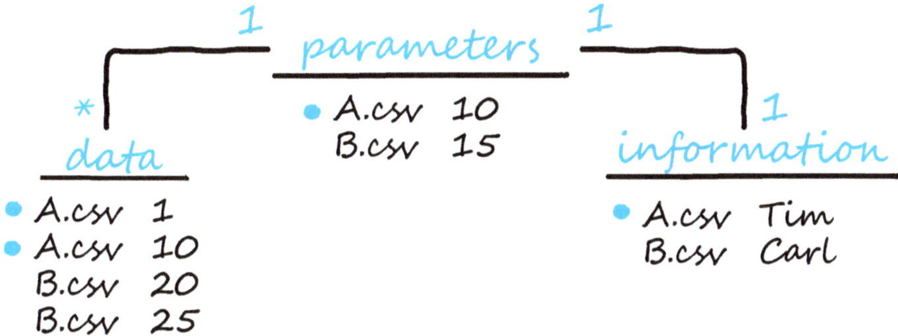

Figure 6.7: Types of connections between the tables *data*, *information* and *parameters* as used herein. The table *parameters* is connected to *data* via a one-to-many relationship and to *information* via a one-to-one relationship. Furthermore, there is a many-to-one relationship between *data* and *information* (not indicated in the figure) regarding the highlighted identity columns.

A *foreign key* is a reference from a row in one table to a row in another table. Reference is made to the primary key in the "other" table. The table with the primary key is referred to as the referenced table or *parent table*; the table with the foreign key is called the *child table*.

Listing 6.10: create_sqlite.py (continued)

```
# %% define tables "data", "information" and "parameters"

from sqlalchemy import Column

# define two column table "parameters"
table_parameters = sqlalchemy.Table(
    "parameters",  # table name
    metadata,  # corresponding metadata
```

```
    Column("cmc_g_l", sqlalchemy.Float),
    Column("id", sqlalchemy.Unicode(255), primary_key=True)
    )
# create the table if not exist
table_parameters.create(checkfirst=True)

# define columns of table "information"
table_information = sqlalchemy.Table(
    "information",  # table name
    metadata,  # corresponding metadata
    Column("Device", sqlalchemy.Unicode(255)),
    Column("Device_ID", sqlalchemy.Unicode(255)),
    Column("Experimental_temperature_degree_C]", sqlalchemy.Float),
    Column("File_processed_by", sqlalchemy.Unicode(255)),
    Column("File_processed_on", sqlalchemy.DateTime),
    Column("Measurement_performed_on", sqlalchemy.DateTime),
    Column("Operator", sqlalchemy.Unicode(255)),
    Column("Sample", sqlalchemy.Unicode(255)),
    Column("Solvent_medium", sqlalchemy.Unicode(255)),
    Column("id", sqlalchemy.Unicode(255),
           sqlalchemy.ForeignKey("parameters.id"))
    )
# create the table if not exist
table_information.create(checkfirst=True)

# define three column table "data"
table_data = sqlalchemy.Table(
    "data",  # table name
    metadata,  # corresponding metadata
    Column("concentration_g_l", sqlalchemy.Float),
    Column("surface_tension_mN_m", sqlalchemy.Float),
    Column("id", sqlalchemy.Unicode(255),
           sqlalchemy.ForeignKey("parameters.id"))
    )
# create the table if not exist
table_data.create(checkfirst=True)
```

In order to check what we achieved so far, we will inspect the created tables. Therefore, we will use the inspect-function of SQLAlchemy. According to Listing 6.11, we specify the engine as the function argument and use the Inspector's method get_table_names to do exactly that.

Listing 6.11: create_sqlite.py (continued)

```
# %% inspect table

# define "inspector"
```

```
inspector = sqlalchemy.inspect(engine)

# get table names
table_names = inspector.get_table_names()

# information
print("\nTables in sqlite-database:")
# loop table names
for _table in table_names:
    # print
    print("   -", _table)
```

We are provided with a list of the just-created tables named: *data*, *information* and *parameters*. The corresponding `Table` objects obtained from the `SQLAlchemy` package are called `table_data`, `table_information` and `table_parameters`, respectively.

6.8 Filling the tables

Obviously, the next question to be addressed is *How do we get data in there*? First of all, we will need to read the respective *csv*-file and subsequently write to the corresponding table of our SQLite-database via an appropriate SQL-type INSERT-statement. Therefore, we define the INSERT-statement with the desired values with the help of `SQLAlchemy`. Only then can we EXECUTE the statement, i. e., write it to the database. Herein, the `iterrows`-method of the obtained `DataFrame` is used to iterate or loop over all rows. This iterator returns the index as an integer and the row as a `pd.Series`. Even though not specifically explained so far, we have been using this type already: A column of a `pd.DataFrame` is a `pd.Series`.

Listing 6.12: create_sqlite.py (continued)

```
# %% get and insert "PARAMETERS"

import pandas as pd

# read parameters file
parameters = pd.read_csv("_PARAMETERS.csv")

# loop rows:
for _idx, _row in parameters.iterrows():
    # info
    print(_row)

    # define insert statement into sqlalchemy.sql.schema.Table
```

```
    # named "table_parameters"
    insert_parameters = table_parameters.insert(values={
            "id"        : _row["id"],
            "cmc_g_l"   : _row["cmc_g_l"]
            })
    # execute insert statement
    insert_parameters.execute()
```

The next question you might be interested in at this point is: *How do I know writing to the database worked out as intended?* That's a good question to ask, and there are basically two indications of success. First, you will not hear any complaints from your script, i. e., there are no error and/or warning messages. Second, we could take a look at the contents of the just written *parameters* table, `table_parameters`, of our SQLite-database. In order to go for the contents of the latter table, we use the `read_sql_table`-function of `pandas` and specify both table name and connection according to Listing 6.13. Piece of cake.

Listing 6.13: create_sqlite.py (continued)

```
# check contents of "parameters" table via pandas
parameters_from_sqlite = pd.read_sql_table("parameters", engine)
# print table
print(parameters_from_sqlite)
```

The console yields the following output:

```
cmc\_g\_l␣id
0␣0.087851␣Polysorbate20.csv
1␣0.088993␣Polysorbate40.csv
2␣0.052587␣Polysorbate60.csv
3␣0.040683␣Polysorbate80.csv
4␣0.054204␣Polysorbate85.csv
```

With this basic understanding in mind, we analogously transfer the contents of the *data* and *information csv*-files *_DATA.csv* and *_INFORMATION.csv* to the corresponding tables of our SQLite-database. As shown in Listing 6.14, we can be more concise compared to Listing 6.12. Instead of explicitly specifying which row value to map to which column of our table, we just use `values=_row` for transferring our *data* to the `data` table of our *surface_tension.db* database. This is possible here because the same column names in both the original *csv*-file and the SQLite-database table were used. Additionally, we use *method chaining* for transferring our *information*. Rather than defining the INSERT-statement, storing it to a variable and finally executing it, we define and execute it in one go. No definition of additional variables is needed in this way.

Method chaining is a programmatic style of invoking multiple method calls sequentially, with each call performing an action on the same object and returning it. Your immediate benefits are less naming of variables and increased readability of your code.

Listing 6.14: create_sqlite.py (continued)

```
# %% get and insert "DATA" and "INFORMATION"

# read data file
data = pd.read_csv("_DATA.csv")

# loop rows:
for _idx, _row in data.iterrows():
    # define insert statement
    insert_data = table_data.insert(values=_row)
    # execute insert statement
    insert_data.execute()


# read information file
information = pd.read_csv("_INFORMATION.csv")
# reshape read DataFrame
information = information.pivot(
        index="id",
        values="value",
        columns="parameter"
        )
# use index column as "regular column"
information = information.reset_index(level=0)
# ensure datetime format
for _c in ["File processed on", "Measurement performed on"]:
    # convert to datetime
    information[_c] = pd.to_datetime(information[_c])
# rename columns to match names of "table_information" (sqlalchemy)
information.columns = [i.replace(" ", "_")
                       for i in information.columns]

# loop rows:
for _idx, _row in information.iterrows():
    # define and execute insert statement ("method chaining")
    table_information.insert(values=_row).execute()
```

6.9 Benefits of a relational structure

At this stage you could—and should—ask yourself: *What is the benefit of moving from our three file csv-structure to the new three table SQLite-structure?* Making use of the relations between the tables, we now can easily answer questions and requests such as *Show me the experimental data of all samples for which the determined CMC is below a certain value*. In Listing 6.15, the CMC of $0.08\,\mathrm{g\,L^{-1}}$ is taken as this exemplary threshold value. Therefore, we use the SELECT-function of `sqlalchemy.sql` and specify that we want to get the entire *data* table `table_data` and the column `cmc_g_l` from the *parameters* table `table_parameters`. We further specify via the WHERE-clause that we want to restrict the latter query to cases with the same `id` column and our actual search condition related to the threshold CMC value.

Listing 6.15: create_sqlite.py (continued)

```
# %% get selected data
#

# define query
query = sqlalchemy.sql.select([ # what do we want?
                                table_data, # all columns
                                table_parameters.c.cmc_g_l  # CMC only
                                ]).where(
                sqlalchemy.sql.and_( # which conditions should apply?
                    table_data.c.id == table_parameters.c.id,
                    table_parameters.c.cmc_g_l <= 0.08,
                    )
                )
```

With this query in hand, we can go back to the Python world and execute the query via `pandas`' `read_sql`-function to obtain a `DataFrame` as shown in Listing 6.16. This allows visualizing the results in just a few lines of code. In this example, joint use of `matplotlib` and `seaborn` is demonstrated.

> *seaborn* is a Python data visualization library based on *matplotlib*. It provides a high-level interface for creating attractive and informative statistical graphics [6]. In other words, it makes a lot things right out of the box. To get started, the `seaborn` homepage also provides a rich gallery of useful examples.[9]

Listing 6.16: create_sqlite.py (continued)

```
# get DataFrame corresponding to query
query_results = pd.read_sql(query, engine)
```

9 https://seaborn.pydata.org/index.html

```
# show
print(query_results)
# concentration_g_l  surface_tension_mN_m id cmc_g_l
# 0 0.004732 56.57  Polysorbate60.csv  0.052587
# 1 0.004732 56.57  Polysorbate60.csv  0.052587
# 2 0.004732 56.57  Polysorbate60.csv  0.052587
# 3 0.004732 56.57  Polysorbate60.csv  0.052587
# 4 0.004732 56.57  Polysorbate60.csv  0.052587
# .. ... ... ... ...

# another plotting module
import seaborn as sns
import matplotlib.pyplot as plt

# plot
sns.lineplot(
    x="concentration_g_l",
    y="surface_tension_mN_m",
    hue="id",
    palette="Paired",
    data=query_results,
    marker="o"
    )

# set labels
plt.xlabel("Concentration / [g/l]")
plt.ylabel("Surface tension / [mN/m]")
# get axes
ax = plt.gca()
# set axes scaling
ax.set_xscale("log")

# add vertical guide to the eye at each distinct CMCs
for _cmc in query_results.cmc_g_l.unique():
    ax.axvline(
        _cmc,
        color="black",
        linestyle="--",
        linewidth=0.5,
        alpha=0.3  # opacity
        )

# save
plt.savefig(
    "selected_sns_plot.png",
    bbox_inches="tight",
    dpi=300
    )
```

The resulting figure to be used, e. g., in a presentation, is shown in Figure 6.8.

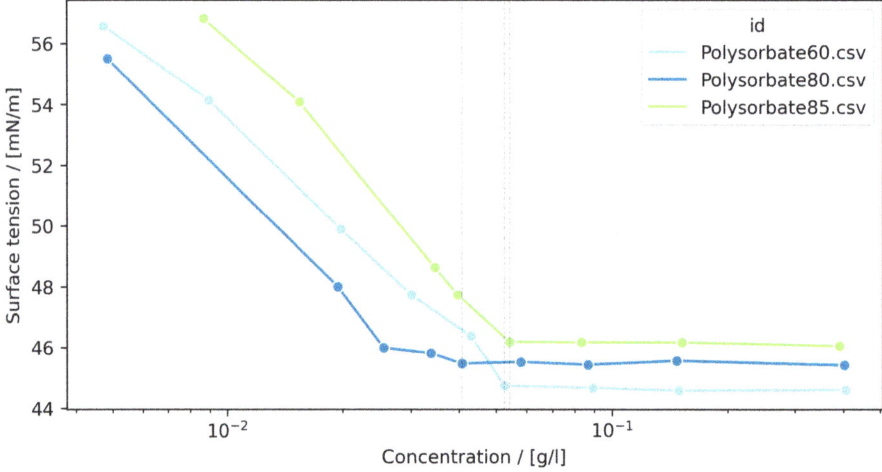

Figure 6.8: Surface tension isotherms (*data* level) filtered for a specific CMC-condition (*information* level). Only *data*-sets meeting the *information* criterion of a CMC-value below 0.08 g L^{-1} are shown.

Similarly, we can select and also show our experimental *data* based on some *information* criterion. In Listing 6.17, the selection of *data* by operator is shown. Therein, we first use the `join`-method of our tables `table_data`, `table_information` and `table_parameters`. Also in this step, we leverage the previously defined relations between the tables.

Listing 6.17: create_sqlite.py (continued)

```
# %% join data and information

# build join
j = table_data.join(table_parameters).join(table_information)

# build query to select specific columns from joined tables
query = sqlalchemy.sql.select([
        table_data,  # all columns from table "data"
        table_information.c.Operator  # "value" column from
                                        # "information"
        ]).select_from(j)

# get corresponding DataFrame
query_results = pd.read_sql(query, engine)
# show
print(query_results)
```

```

# plot
sns.lineplot(
    data=query_results,
    x="concentration_g_l",
    y="surface_tension_mN_m",
    hue="Operator",  # color by operator
    style="id",  # unique linestyle for each sample
    palette="Paired",
    markers=True,
    units="id",  # what is an "individual" plot
    estimator=None,  # no aggregation as mean, etc.
    )

# set labels
plt.xlabel("Concentration / [g/l]")
plt.ylabel("Surface tension / [mN/m]")

# get axes
ax = plt.gca()
# set axes scaling
ax.set_xscale("log")

# save
plt.savefig(
    "data_by_operator_sns_plot.png",
    bbox_inches="tight",
    dpi=300
    )
```

Plotting the query results provides an overview of operators who conducted the measurements as shown in Figure 6.9. In summary, we are now able to seamlessly navigate between the levels of *data*, *information* and *parameters*. This will be helpful in many questions you might have about the exemplary dataset. Of course, this becomes even more interesting if it's really about *your* data.

6.10 Wrap up

This chapter introduced ways of storing *data*, *information* and derived *parameters* on different levels of complexity:
- organized folder structure and a
- SQLite-database.

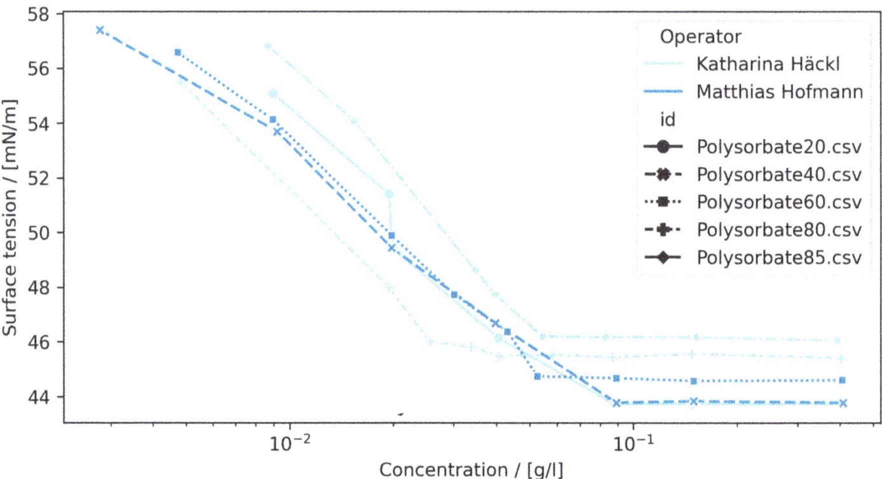

Figure 6.9: Experimental surface tension isotherms (*data* level) highlighted by a operator (*information* level). All datasets are shown and colour-coded by the operator. Furthermore, the measurements are individualized by line style and marker.

Using `SQLAlchemy` for interfacing with the herein presented SQLite-database equally enables connecting to MySQL or PostgreSQL-databases—if those are available in your institution. Upon going through the experimental raw data files, adapting the so-far developed module `data_to_information` was demonstrated in response to identified parameter extraction difficulties for certain files by taking into account the additional keyword `gradient_threshold`. Creation of the SQLite-database *surface_tension.db* and the relevant tables, `table_data`, `table_information` and `table_parameters` therein, was followed by writing the already collected values to this structure.

The chapter concludes with accessing and filtering *data*, *information* and derived *parameters* across these levels for the purpose of visualization according to a specific request. In this step, the relationships between the tables are leveraged by generating queries via `SQLAlchemy` and collecting the corresponding data through the `read_sql`-function of `pandas`.

7 How to visualize data and information

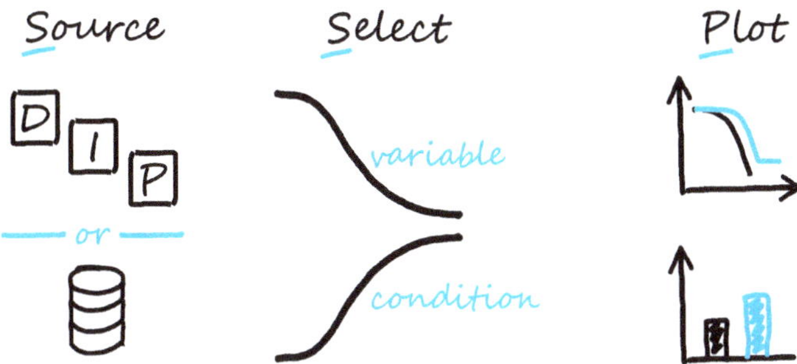

Once we have stored experimental results in an organized folder structure, summary file or database as outlined in Chapter 6, we arrive at yet another starting point for a whole new field: *data visualization*. Also referred to as *data viz* (if you want to search for it online), countless books have been written on what good or bad visualization means in a specific context. Also, you will stumble across the term *storytelling with data*, which I consider a notion that provides much insight into the actual goal of this process step. Therefore, I will not engage in the "religious wars" but rather merely provide you with some lessons I have personally experienced to be beneficial in my daily work:

- Look at your data in light of the question you would like to answer.
- Develop a working familiarity with design, statistics and computer science since their intersection is the basis of modern data visualization.
- Make the key message of your visualizations easy to understand—both for your audience and yourself.
- Do not overload a visualization. Less might indeed be more.
- Know your data—and make use of this!

7.1 Source and visualization options

As in the previous chapters, I would like to introduce multiple options for visualizing our so-far collected *data*, *information* and *parameters*. Once again, they exhibit diverse levels of user friendliness and complexity. We will take a look at the following options:

- Python script + packages `matplotlib` and/or `seaborn`, and
- drag and drop solutions, and dedicated software (such as Microsoft® Power BI Desktop®, Tableau).

https://doi.org/10.1515/9783110788433-007

Before diving into the visualization part, I would like to point out the various options for accessing all of the so far collected data. We can either use the contents of the three individual *csv*-files or directly access our SQLite-database as indicated in Figure 7.1.

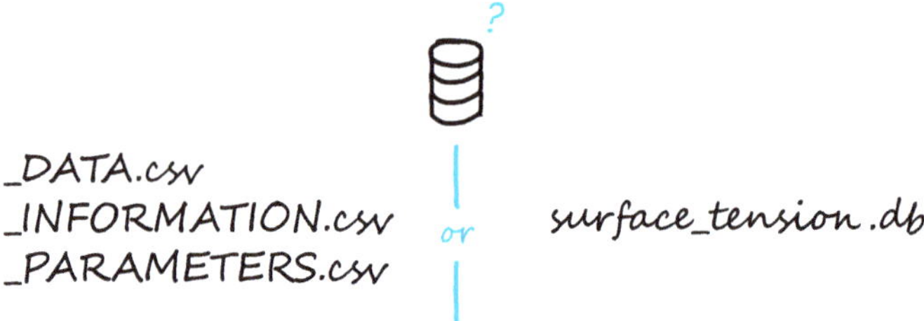

Figure 7.1: Data-source options. We can either draw *data*, *information* and *parameters* from the individual *csv*-files or from the previously generated SQLite-database holding this very content in three tables, respectively.

For clarity and depending on the need of your actual use case, I will show both access options to highlight their validity for our visualization task. In the end, it comes down to personal preference and the amount of data to be expected overall. As there is some structure in either of the cases, the transition to a database solution, starting from the file-based approach, can be readily achieved at a later point in time.

Generally, you will not be interested in plotting *all* the available data but merely a certain selection based on a criterion relevant to answer your specific question. Also for this step, I will present two methods relying mostly on either `pandas` or `SQLAlchemy` as suggested by Figure 7.2. It is almost certain that your current questions will differ from those that arise tomorrow.

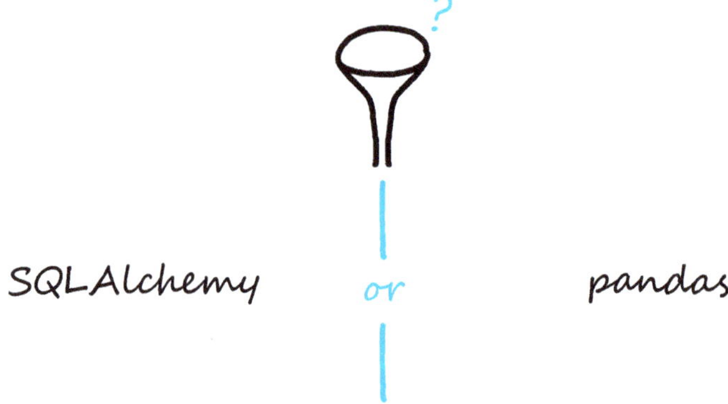

Figure 7.2: Data select options: *data*, *information* and *parameters* can be boiled down using either `SQLAlchemy` or `pandas` (or a combination of both).

Also for the plotting part, two possibilities will be introduced (see Figure 7.3). On the one hand, there's `seaborn` coming with lots of functionality of the box—under the condition that our `DataFrames` are formatted “properly”. On the other hand, we can use `matplotlib` if we are willing to invest some additional time in writing a few more lines of code with the benefit of increased flexibility. Of course, we can also do some basic plotting using `seaborn` and fine-tuning with the help of `matplotlib`.[1]

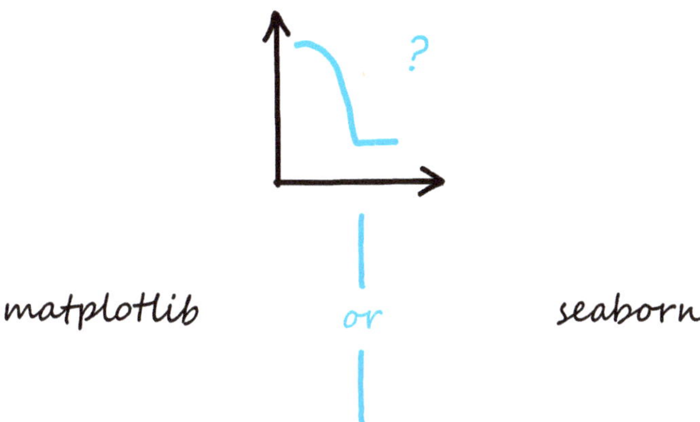

Figure 7.3: Plotting options: Out of the box convenience of `seaborn` or `matplotlib`.

An overview of the available source, selection and plotting options is given in Figure 7.4. In the following sections, an example for each of the available pathways will be given to demonstrate the basic ideas. Of course, the “best” combination may depend on your preferences and specific use case.

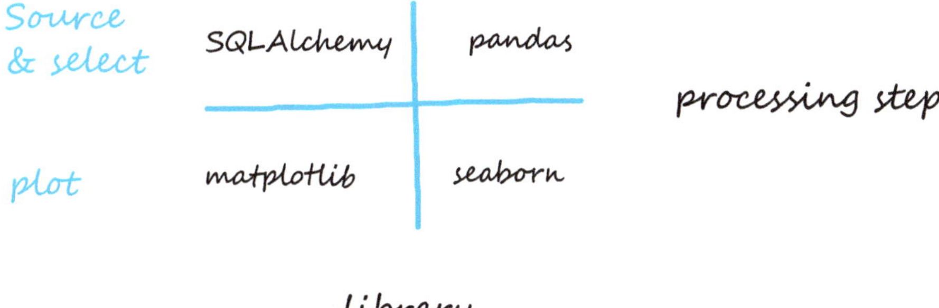

Figure 7.4: As for the plotting step, there are two options for sourcing and selecting *data*, and *information*: `pandas` and `SQLAlchemy`. Mix and combine as you like.

---

1 Other frequently used plotting libraries are `bokeh` (see https://docs.bokeh.org) and `plotly` (see https://plotly.com).

7.2 Python script + matplotlib and/or seaborn

7.2.1 Sourcing and selection via SQLAlchemy

To get hands on our *data*, *information* and *parameters* tables, we first have to connect to our *surface_tension.db* database. As in the previous code snippets, we will use a `MetaData` object to gain further insight on our database—just in case we have forgotten since the previous pages. In particular, we define an `Inspector` to get the names of our tables according to Listing 7.1.

Listing 7.1: sqlalchemy_matplotlib_way.py

```
# module for handling sqlite
import sqlalchemy

# create connection to on disk database "surface_tension.db"
engine = sqlalchemy.create_engine(
            'sqlite:///surface_tension.db',
            echo=True
            )

# create bound metadata
metadata = sqlalchemy.MetaData(engine)
# load all available table definitions from the database.
metadata.reflect()

# define "inspector"
inspector = sqlalchemy.inspect(engine)

# show available tables
for _table_name in inspector.get_table_names():
    # show this table
    print("  -", _table_name)
```

The console yields the following output:

```
␣-␣data
␣-␣information
␣-␣parameters
```

To further work with these tables, we have to access them individually and finally JOIN them by making use of the previously defined relations, i. e., connections between them as shown in Listing 7.2.

Listing 7.2: sqlalchemy_matplotlib_way.py (continued)

```
# %% get tables
#

# access database tables
table_data = metadata.tables["data"]
table_information = metadata.tables["information"]
table_parameters = metadata.tables["parameters"]

# join results
join = table_data\
        .join(table_parameters)\
        .join(table_information)
```

Next, we proceed with the selection of both *data* and the extracted *parameters* based on a criterion on the *information* level as shown in Listing 7.3. In our example: *What data and associated critical micellar concentration (CMC) values have been measured prior to January 28, 2022 before 12 AM?* In order to arrive at the relevant query, we start with the SELECT-statement, making use of the previously defined JOIN and boil that down further via the WHERE-clause. Within this lengthy statement, you can consider the dots before `select_from` and `where` as separators. They are markers of method chaining.

Run the code shown in Listing 7.3 after commenting/uncommenting the restrictions introduced via the `select_from` and `where` methods to get a feeling of the applied method chaining. It makes possible writing complex queries in a concise form.

Listing 7.3: sqlalchemy_matplotlib_way.py (continued)

```
# specify sqlalchemy query
from sqlalchemy.sql import select
import datetime

query = select([
                table_data,  # get all data columns
                table_parameters.c.cmc_g_l  # get CMC column
                ])\
            .select_from(join)\
                .where(
                    table_information.c.Measurement_performed_on\
                        <= datetime.datetime(2022, 1, 28, 12, 0, 0)
                    )
```

To prepare ourselves for the plotting task, we rely once more on pandas' read_sql-function to obtain the query results as a DataFrame according to Listing 7.4.

Listing 7.4: sqlalchemy_matplotlib_way.py (continued)

```
# import for DataFrame handling
import pandas as pd

# get data corresponding to query
data = pd.read_sql(query, engine)
# show
print(data)

# === Console output: ===
#      concentration_g_l  surface_tension_mN_m id cmc_g_l
# 0 0.004732 56.57  Polysorbate60.csv  0.052587
# 1 0.008944 54.14  Polysorbate60.csv  0.052587
# 2 0.019793 49.89  Polysorbate60.csv  0.052587
# 3 0.030056 47.73  Polysorbate60.csv  0.052587
# 4 0.043003 46.37  Polysorbate60.csv  0.052587
# 5 0.052587 44.74  Polysorbate60.csv  0.052587
# 6 0.088993 44.67  Polysorbate60.csv  0.052587
# 7 0.149110 44.58  Polysorbate60.csv  0.052587
# 8 0.406020 44.60  Polysorbate60.csv  0.052587
# 9 0.004837 55.50  Polysorbate80.csv  0.040683
# 10 0.019440 48.00  Polysorbate80.csv  0.040683
# 11 0.025564 45.99  Polysorbate80.csv  0.040683
# 12 0.033768 45.81  Polysorbate80.csv  0.040683
# 13 0.040683 45.47  Polysorbate80.csv  0.040683
# 14 0.057990 45.52  Polysorbate80.csv  0.040683
# 15 0.086586 45.43  Polysorbate80.csv  0.040683
# 16 0.147310 45.56  Polysorbate80.csv  0.040683
# 17 0.402270 45.41  Polysorbate80.csv  0.040683
```

Starting from this DataFrame, we can start our plotting as outlined in Subsections 7.2.3 or 7.2.4.

7.2.2 Sourcing and selection via pandas

Collecting and selecting data to answer a specific question is also possible via pandas, starting from the previously defined "summary" *csv*-files holding *data*, *information* and *parameters*. In this example, we are looking for CMC-values and the corresponding *data* for which a minimum number of concentrations has been studied. This can be understood as some kind of quality criterion of the overall measurement: Only if the concentration spacing between our measured points is close enough, can we reliably extract a CMC-value from the experimental *data*. One abstraction level higher, we select *data* and *parameters* based on a condition on the raw *data*. In terms of the *data pyramid*, the level of *information* is not involved in this specific question. Interestingly, the required Structured Query Language (SQL)-like JOINs can be fully performed between DataFrames via the merge-function of pandas. Finally, we will do the plotting. Once more, a piece of corresponding pseudo-code could take the form as shown in Listing 7.5.

Listing 7.5: pandas_seaborn_pseudocode.py

```
# %% 1) read csv files

# %% 2) select relevant "data"

# %% 3) join to "parameters"

# %% 4) visualize
```

In the first steps shown in Listing 7.6, we will use the read_csv-function and the groupby-method for the selection of *data* . Herein, we will restrict our analysis to samples for which a minimum number of eight concentration–surface tension value pairs have been collected. The variable number_of_observations is a pd.Series obtained by grouping our data by the sample's id, counting the number of entries and finally selecting the first column of the resulting DataFrame via the iloc-method. This is another example of the powerful method chaining in Python. The selection of rows of interest is referred to as *subsetting*. In our example, the condition is specified via number_of_observations >= 8. This returns a pd.Series of True and False. What's left to do is to pack this condition in square brackets of our number_of_observations pd.Series, get the remaining index values and convert those into a list representing the sample names meeting the specified condition.

Listing 7.6: pandas_seaborn_way.py

```
# %% 1) read csv files

# import for data handling
import pandas as pd

# read data
data = pd.read_csv("_DATA.csv")
# read parameters
parameters = pd.read_csv("_PARAMETERS.csv")


# %% 2) select relevant "data"

```

```
# define selection criterion: number of observations per sample
number_of_observations = data.groupby(by="id").count().iloc[:,0]

# get names of samples meeting our criterion / "subsetting"
target_samples = number_of_observations[
                    number_of_observations >= 8
                    ].index.to_list()

# print names of samples meeting the criterion
print("TARGET SAMPLES:")
for _s in target_samples:
    print("  -", _s)
```

With this list of target samples, we can boil down our *data* of interest and merge the corresponding *parameters* according to Listing 7.7. For this purpose, we use the `isin`-method of the `id`-column of our `DataFrame` `data` to check against the previously obtained list of "target samples". Again, the result of this test is a `pd.Series` of `True` and `False` values. Packing these into square brackets of the `data` `DataFrame` narrows it down for further processing.

Listing 7.7: pandas_seaborn_way.py (continued)

```
# %% 3) join to "parameters"

# boil down "data" to relevant part
data = data[data["id"].isin(target_samples)]

# merge "data" and "parameters" to "selection"
selection = pd.merge(
    data,  # "left" DataFrame
    parameters,  # "right" DataFrame
    left_on="id",  # id of left df to merge on
    right_on="id",  # id of right df to merge on
    validate="many_to_one"  # optional!
    )
```

Python is granted some powerful data merging capabilities via the `pandas` `merge`-function. Next to declaring which `DataFrames` are to be merged as left and right tables, we can further specify on which columns the merge is to be carried out via the optional `left_on` and `right_on` keywords. It is also possible to carry out a merge based on two or more columns of the specified left and right `DataFrames`. A typical example for this scenario in the present context of chemical substances are two tables each having a name and a batch column. Merging those `DataFrames` is possible via specification of a list instead of an individual string. A further optional keyword argument to use in the latter code example is validate for checking the desired type of relation between the tables. If the validation condition is not met, a MergeError will be triggered.

Check the behaviour of the code snippet shown in Listing 7.7 in case of omitting the left_on and right_on-keywords, as well as using one of the other options for the validate keyword:
- one_to_one or 1:1: check if merge keys are unique in both left and right datasets.
- one_to_many or 1:m: check if merge keys are unique in left dataset.
- many_to_one or m:1: check if merge keys are unique in right dataset.
- many_to_many or m:m: allowed, but does not result in checks.

The only thing left to do at this stage is visualizing the results filtered according to the previous steps. This example is continued in subsection 7.2.4, wherein the main plotting task is carried out via seaborn, and only some plot setup is taken care of by matplotlib.

7.2.3 Plotting via matplotlib

At the beginning of each and every visualization task, it is a good idea to recall what we actually intend to show. In the best possible case, we exclusively have the data we really need for the targeted representation (see the previous Subsections 7.2.1 and 7.2.2). We can get a quick look at the DataFrame's structure by printing its column names as shown in Listing 7.8 or by typing data.columns in the console once data is available.

Listing 7.8: sqlalchemy_matplotlib_way.py (continued)

```
# %% plotting
#

# show available columns
for _c in data.columns:
    print(_c)
```

The console yields the following output:

```
concentration_g_l
surface_tension_mN_m
id
cmc_g_l
```

To repeat: To obtain an overview of the measured samples before a certain point in time, we want to plot surface tension against concentration on a logarithmic scale for each sample and additionally show the corresponding CMC-values. As indicated by this description, the plotting task again requires some looping. The plotting library `matplotlib`, however, is not capable of this task *alone*. Therefore, we will have to provide the surface tension and concentration *data* and CMC *parameters* corresponding to a specific sample one after another. In this situation, the `DataFrame`'s `groupby`-method is there to help us. In short, it cuts the overall `DataFrame` into smaller `pd.DataFrames` according to our grouping wishes. In our example, we obviously want to group by sample, i. e., the `id` column as shown in Listing 7.9.

Listing 7.9: sqlalchemy_matplotlib_way.py (continued)

```
# import basic plotting library
import matplotlib.pyplot as plt

# loop by sample
for this_sample, this_sample_data in data.groupby(by="id"):
    # info
    print(this_sample)
    print(this_sample_data)
```

The console yields the following output:

```
Polysorbate60.csv
concentration_g_l␣surface_tension_mN_m␣␣␣␣␣␣␣␣␣␣␣␣␣␣␣␣id␣␣cmc_g_l
0␣␣␣␣␣␣␣␣0.004732␣␣␣␣␣␣␣␣␣␣␣␣␣␣␣␣56.57␣Polysorbate60.csv␣0.052587
1␣␣␣␣␣␣␣␣0.008944␣␣␣␣␣␣␣␣␣␣␣␣␣␣␣␣54.14␣Polysorbate60.csv␣0.052587
2␣␣␣␣␣␣␣␣0.019793␣␣␣␣␣␣␣␣␣␣␣␣␣␣␣␣49.89␣Polysorbate60.csv␣0.052587
3␣␣␣␣␣␣␣␣0.030056␣␣␣␣␣␣␣␣␣␣␣␣␣␣␣␣47.73␣Polysorbate60.csv␣0.052587
...
```

After this grouping step, we have all the *data* we need to plot the results corresponding to one sample contained in the `DataFrame` named `this_sample_data`. The next step to be carried out is plotting the experimental data and the extracted CMC. The final touches for completing the plot are, among some others, affixing appropriate labels and a title.

If you are trying to carry out this plotting procedure by yourself, rewriting the above recipe in the form of a pseudo-code is a helpful exercise.

The code carrying out these steps is given in Listing 7.10, the resulting visualization is shown in Figure 7.5.

Listing 7.10: sqlalchemy_matplotlib_way.py (continued)

```
    # plot experimental data
    plt.plot(
        this_sample_data["concentration_g_l"],
        this_sample_data["surface_tension_mN_m"],
        marker="o",
        label=this_sample,  # legend entry
        )

    # cmc indication line
    plt.axvline(
        this_sample_data["cmc_g_l"].to_list()[0],
        color="black",
        linewidth=0.5
        )

# set axis scaling
plt.xscale("log")

# add legend (filled via "label=" in plot-function calls)
plt.legend()

# figure "cosmetics"
plt.xlabel("Concentration / [g/l]", size=14)
plt.ylabel("Surface tension / [mN/m]", size=14)
plt.title("Surface tension isotherms measured before {}".format(
    datetime.datetime(2022, 1, 28, 12, 0, 0),
    ),
    size=20
    )

# save
plt.savefig(
    "sqlalchemy_matplotlib_way.png",
    bbox_inches="tight", # removes extra white space around the figure
    dpi=300  # dots per inch (image quality)
    )
```

7.2.4 Plotting via seaborn

As mentioned in this chapter's introduction, the seaborn plotting library allows visualization of large datasets—if properly formatted—requiring only a few lines of code. Keep in mind that plotting via seaborn tends to be slower compared to visualizations using matplotlib. In this example, we will do a side-by-side plot of *data*, i. e., our surface tension isotherms and *parameters*, i. e., the extracted CMC-values from the raw data as a bar plot. The basic setup of this visualization is achieved via matplotlib: We define a plot layout and the overall size as shown in Listing 7.11.

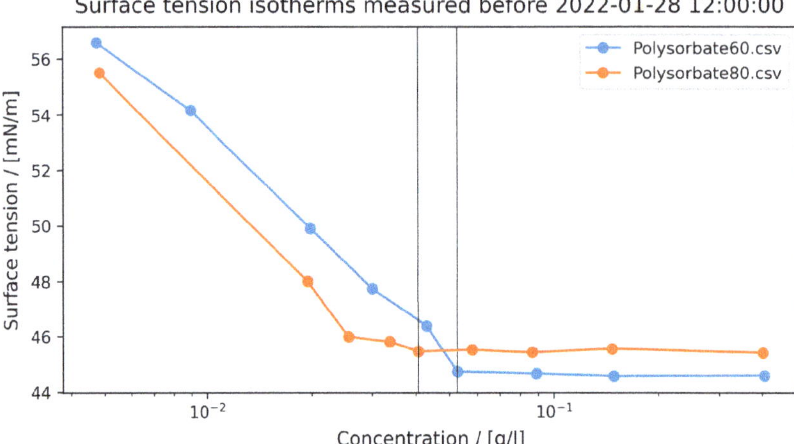

Figure 7.5: Surface tension isotherms (*data*) and corresponding CMC-values (*parameters*) filtered according to a condition (surface tension isotherm measured before 12:00 of January 28, 2022) on the *information* level.

Listing 7.11: pandas_seaborn_way.py (continued)

```
# %% 4) visualize

# import plotting library
import seaborn as sns
import matplotlib.pyplot as plt

# define 1 x 2 plot grid
fig, (ax1, ax2) = plt.subplots(1, 2)
# set figure size
fig.set_size_inches(8, 4)
```

For performing the actual plotting task, we set the obtained variables `ax1` and `ax2` of type `matplotlib.axes._subplots.AxesSubplot` as arguments in the `seaborn` plotting functions. The corresponding code is shown in Listing 7.12.

Listing 7.12: pandas_seaborn_way.py (continued)

```
# barplot of CMCs ("parameter" level)
sns.barplot(
    data=selection,
    y="id",  # use "id" column value as y-axis --> horizontal bars
    x="cmc_g_l",
    palette="Blues",
    ax=ax1  # place plot in "left" space
```

```
    )

# lineplot of surface tension isotherms ("data" level)
sns.lineplot(
    data=selection,
    x="concentration_g_l",
    y="surface_tension_mN_m",
    hue="id",  # color by "id" column value
    palette="Blues",
    marker="o",
    ax=ax2  # place plot in "right" space
    )
```

Again, we conclude the visualization task by applying some “cosmetic” changes. We set the axis-scaling of the right-hand side surface tension isotherm plot to logarithmic (as this is the convention in the field of surface chemistry) and add a meaningful title. The code of Listing 7.13 is terminated by saving the plot as the graphic shown in Figure 7.6 to a file of a defined resolution.

Listing 7.13: pandas_seaborn_way.py (continued)

```
# set log scaling on x-axis
ax2.set_xscale("log")

# set labels
ax1.set_xlabel("CMC / [g/L]")
ax2.set_xlabel("Concentration / [g/L]")
ax2.set_ylabel("Surface Tension / [mN/m]")

# add common title
plt.suptitle("Surface Tension Isotherm data and CMC values for \
samples measured at >= 8 concentrations")

# hide legend title
plt.legend(title=None)

# adjust suplot sizes to fit them the plot area
fig.tight_layout()

# save figure
fig.savefig(
    "pandas_seaborn_way.png",
    dpi=300
    )
```

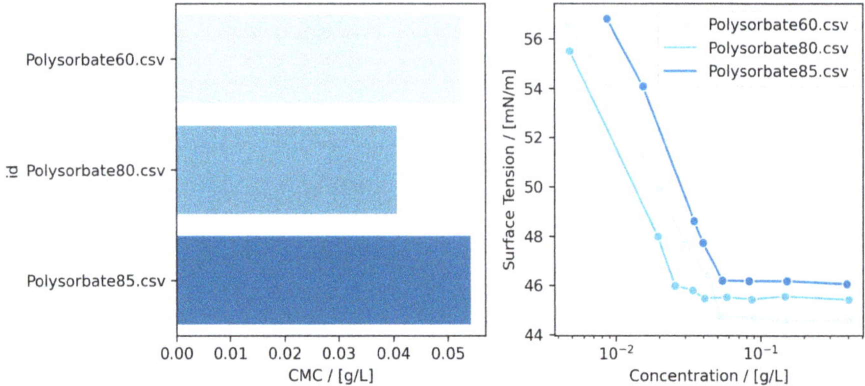

Figure 7.6: CMC values (on level *parameters*) and corresponding surface tension isotherms (on level *data*) filtered according to a condition on the *data* level. Only surface tension isotherms for which at least eight surface tension concentration pairs have been collected are shown. Note that *information* (in the sense of *metadata*) is not taken into account herein.

In Figure 7.6, a horizontal bar plot is used to visualize the CMC-values corresponding to the samples. There are some advocates for using them over their vertical counterparts. Most importantly, reading the ticks is easier. Rotating your head, hardcopy or display is not necessary.

Build the corresponding vertical bar-plot counterpart to Figure 7.6 using either the `x`- and `y`-keywords and/or the `orient`-keyword of the `seaborn.barplot`-function (Check this function's documentation by typing `sns.barplot?` in the console.).

7.3 Dedicated visualization software

An alternative approach to visualize your results for presentation—and in particular for exploratory purposes—is interactive data visualization. This means that you need to be able to "slice and dice" your *data* in essentially no time to test your hypotheses at short notice. In this regard, there are at least two outstanding software solutions: *tableau*[2] and *Microsoft® Power BI Desktop®*,[3] but there are certainly many more alternative *business intelligence* programs out there, which are even free of charge.

In the following, I will show the basics of using our so-far created SQLite-database and one additional data source in Microsoft® Power BI Desktop®. This tool is both free, and as a Microsoft® Office® user you will find your way therein quite soon.

2 https://www.tableau.com/

3 https://powerbi.microsoft.com/de-de/

Because Microsoft® Power BI Desktop® *can* be used to a large extent without coding for our herein presented purpose (visualizing and filtering experimental *data*, *information* and *parameters* across these levels), I will not include the pseudo-code part from the previous sections, but just provide a list of steps to be carried out:

- connect to one or more additional data sources, e. g., *csv*-files, *xlsx*-files, SQLite-databases and many, many more,
- create and/or check the connections between the data sources (if they exist),
- select and specify visuals and
- refresh the underlying collected *data*, *information* and *parameters* if changing these data sources.

7.3.1 Connecting to data via Microsoft® Power BI Desktop®

First of all, we need to make our SQLite-database available to Microsoft® Power BI Desktop®. This is possible via *Open Database Connectivity (ODBC)*, a standard for database management systems as shown in Figure 7.7. If not already available on your machine, you will have to download and install an ODBC driver as a translation layer between Microsoft® Power BI Desktop® and the SQLite-database from the web (see Appendix F).

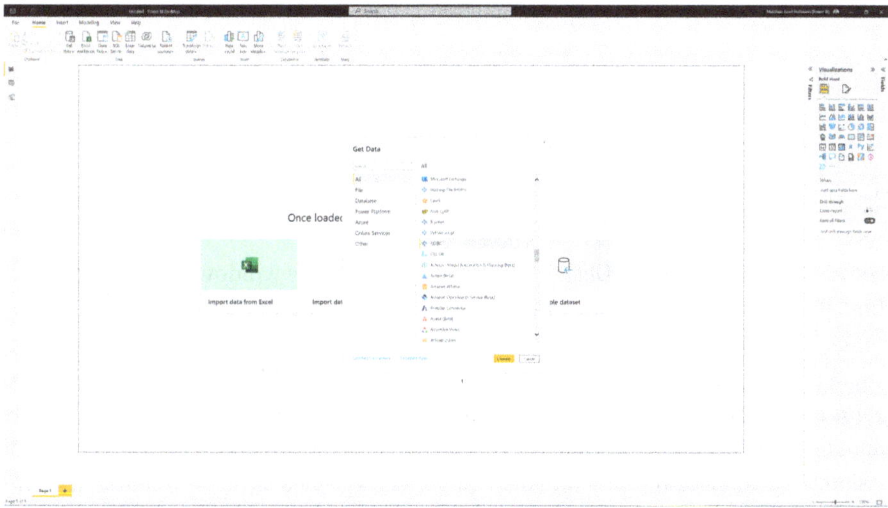

Figure 7.7: Data connection options in Microsoft® Power BI Desktop®. We use the ODBC-connector to get access to our previously created and filled SQLite-database.

Once the type of connector is specified as ODBC, we will move on to the more precise specification. As Data Source Name (DSN) we select the name from the dropdown menu that we previously specified during installation of the ODBC-driver (see Appendix F). We declare database=<path> as the connection string, whereas <path> represents the actual path to our local SQLite-database. Finally, our ODBC-configuration with specified DSN and connection string might look Figure 7.8.

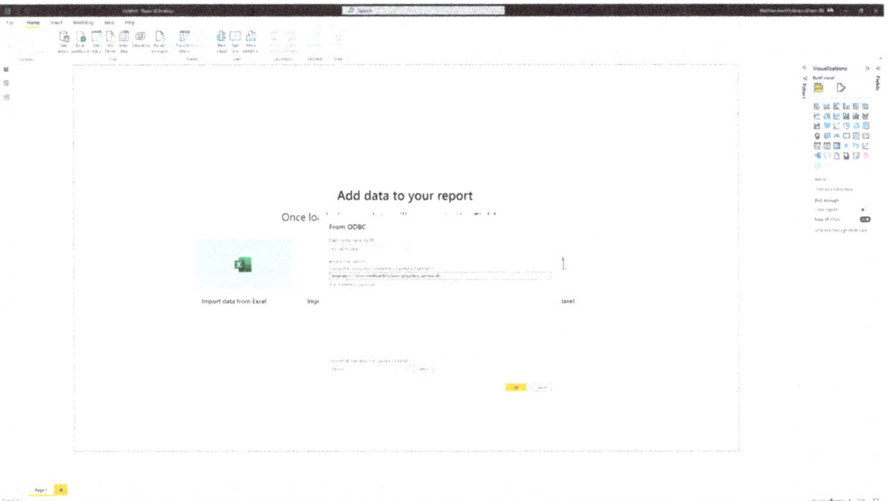

Figure 7.8: ODBC-configuration with specified DSN and connection string for making the SQLite-database *surface_tension.db* accessible to Microsoft® Power BI Desktop®.

Please note that, instead of connecting to the SQLite-database, *data*, *information* and *parameters* could also be sourced from the respective *csv*-files introduced in the previous chapters. Connecting to those *csv*-files individually requires less specification compared to the SQLite-database, but the relationships will have to be defined explicitly within Microsoft® Power BI Desktop®.

Next, Microsoft® Power BI Desktop® opens up a dialogue window showing all available tables in the specified database. In our example, we find the tables *data*, *information* and *parameters* as shown in Figure 7.9. On the right-hand side of the window, a preview of the respective selected table is given, highlighted via a darker background in the list of tables on the left. In our example, the contents are sufficiently small to show the entire tables in the preview area. Interestingly, Microsoft® Power BI Desktop® recognizes the relationships between the tables *data*, *information* and *parameters* previously defined within the SQLite-database using SQLAlchemy. This can be seen from the columns data and information of the selected table parameters holding the value Table in the preview window shown in Figure 7.9. Expanding these tables would “glue” the tables together according to the key specifications made in the previous chapter. Because we would like to use the contents of all three available tables, we hit the checkboxes left to the respective table names and terminate the dialogue

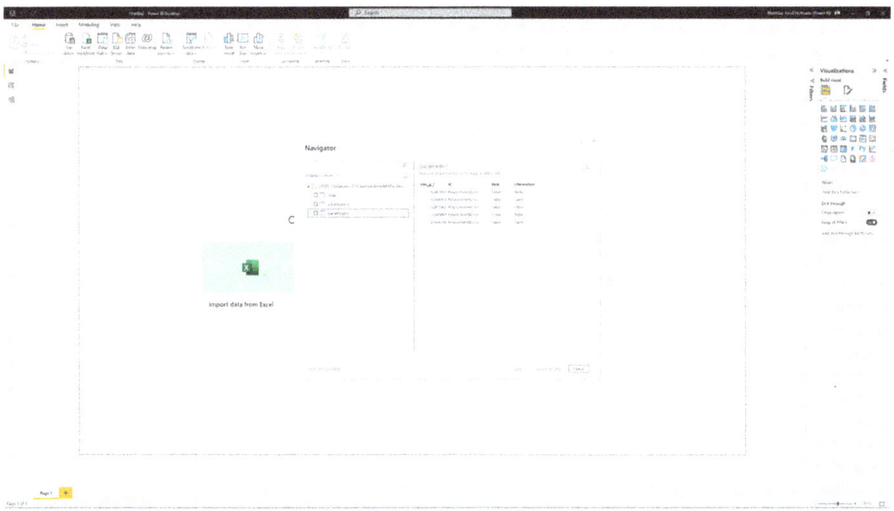

Figure 7.9: Selecting tables contained in the SQLite-database *surface_tension.db* for further use within Microsoft® Power BI Desktop®.

either via the "Load" or the "Transform Data" button on the bottom right-hand side of the window.

As a next step, we want to make sure the relationships between our tables are in agreement with the data model that we initially had in mind and specified in our Python code.

This step is fully redundant and merely serves as another layer of "validity check".

To check the connection between the tables `data`, `information` and `parameters` via the `id` column, we navigate to the "Data Model" view indicated in Figure 7.10 and go to the "Manage Relationships" button on the top right. In this step, we set the connections between the tables to be bidirectional. This means setting a filter on table A also sets a filter on table B and *vice versa*. Furthermore, we hide the `id` columns in the tables `data` and `information` (indicated via the crossed out eye symbol). This has the benefit of making only one specific information set available at only one location. In our case, we take the `id`-information exclusively from the `parameters` table.

On to the next data source: As previously mentioned, we assume for the sake of the example that we obtain additional information on our experimentally characterized samples from a—mystical—external source. In reality, this might be some type of historic data collection you may have unearthed in the course of your project or data collected by colleagues from another department working with a different focus. In our case, the additional data source comes in the form of a Microsoft® Excel®-file with the

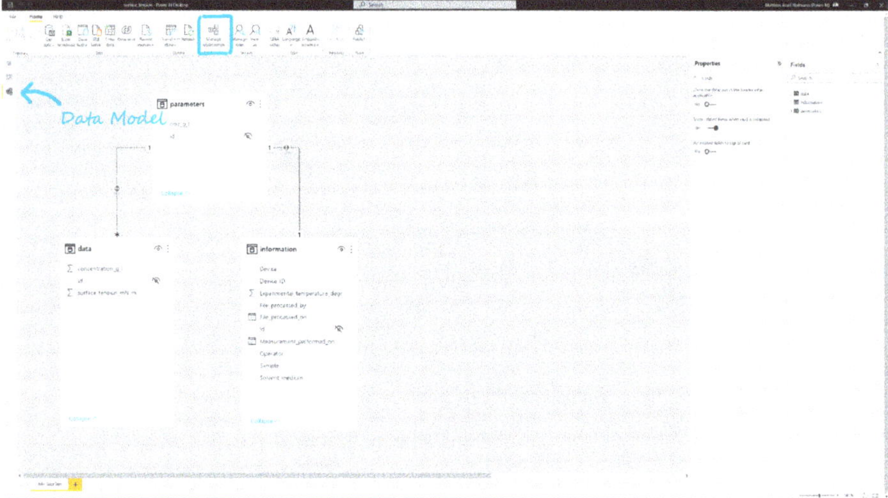

Figure 7.10: Specifying the data model relying on the SQLite-database *surface_tension.db* in Microsoft® Power BI Desktop®. The lines between the indicated tables `data`, `information` and `parameters` highlighted by triangles pointing towards the lines' ends represent a bidirectional connection via the respective `id` columns.

columns `Product`, `CAS`,[4] `Molar Mass` and `Chemical Formula`. Getting the contents of this Microsoft® Excel®-file is as easy as hitting the "Get Data" button, navigating to the file of interest and specifying the target sheet. The dialogue window of Figure 7.11 is widely similar to the previously shown interface appearing in the course of specifying the connection to our SQLite-database in Figure 7.9. Here, also a preview of the sheet's contents is given on the right-hand side.

7.3.2 Building a data model

The—more or less—challenging part for this type of scenario is the mapping of our *data* stored in the SQLite-database *surface_tension.db* to the content of the newly introduced Microsoft® Excel®-file. In this example, we will "rebuild" the `id` column of the SQLite-database in the contents of *Polysorbate.xlsx* to establish a connection between the so-far unrelated tables. In concrete terms, we have to remove the white space from the column `Product` of *Polysorbate.xlsx* and add a *.csv* suffix in order to match the `id`-pattern of the SQLite-tables `data`, `information` and `parameters` according to Figure 7.12.

4 A CAS Registry Number is a unique numerical identifier assigned by the Chemical Abstracts Service assigned to every chemical substance described in the open scientific literature.

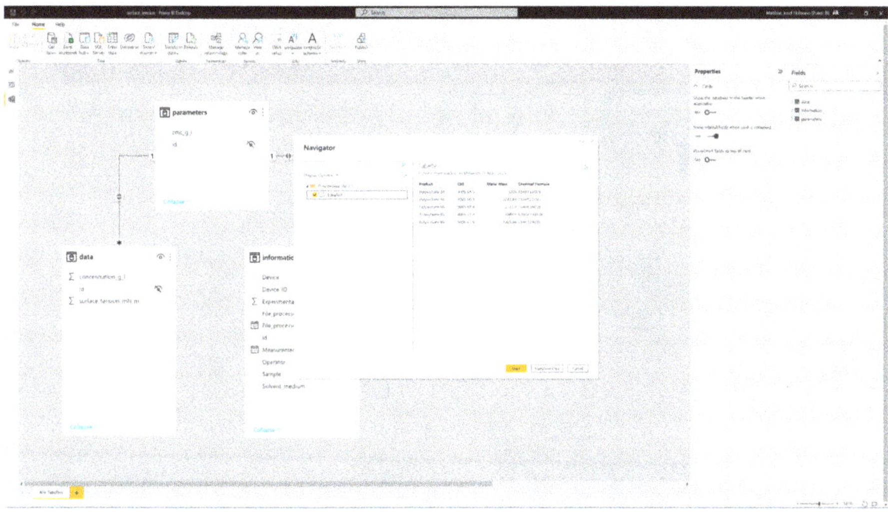

Figure 7.11: Selecting the sheet of interest from the Microsoft® Excel®-file *Polysorbate.xlsx* for further use within Microsoft® Power BI Desktop®.

Parameters.xlsx
└ Product

surface_tension.db
└ id

e.g. Polysorbate 40 ⟶ Polysorbate|40.csv

Figure 7.12: Mapping the `Product` column of *Polysorbate.xlsx* to the `id` column of the SQLite-database *surface_tension.db*: remove the white space before the number and add the suffix *.csv*.

Within Microsoft® Power BI Desktop®, we achieve this via duplicating the original `Product` column of *Polysorbate.xlsx* and applying the latter transformation steps using the tools available therein according to Figure 7.13.

To conclude the model building part including data from both SQLite and the posterior obtained Microsoft® Excel®-file, we will map the newly gained *information* to the already existing `information` table from SQLite via the `id` column generated according to the previous steps. Note that we renamed the table to `information (extended)` within Microsoft® Power Query®. Also for the relationship between `information` and `information (extended)`, we use the bidirectional type as shown in Figure 7.14.

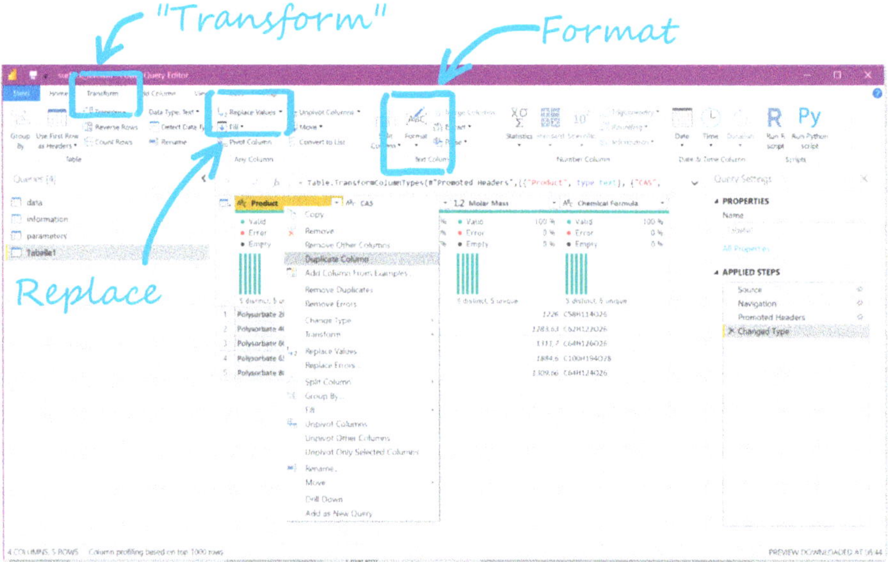

Figure 7.13: Duplication of the original `Product` column for constructing the mapping `id` column from it. Eliminating white spaces is possible via replacing values. Adding suffixes is done using the format functions in the "Transform" section.

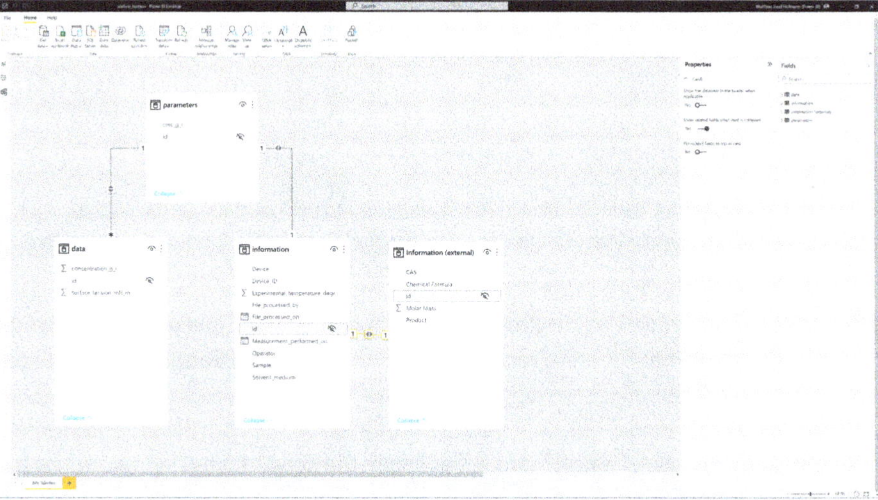

Figure 7.14: Connecting the `information (extended)` table to the already existing table `information` via the derived `id` column using a bidirectional 1:1 relationship.

7.3.3 Visualizing

Now that we are finally in a position to have our data at our fingertips, we can begin with the visualizing part. Therefore, Microsoft® Power BI Desktop® offers a rich

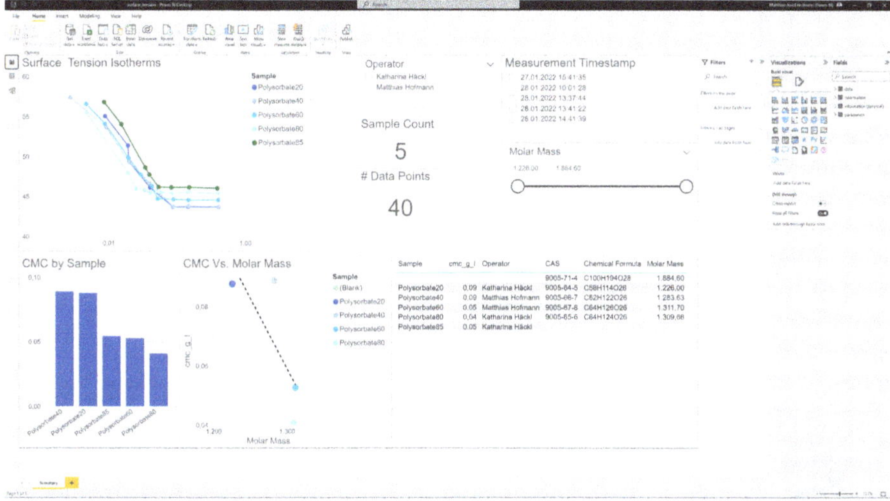

Figure 7.15: Combined visualization of *data* originating from a first data source, the SQLite-database *surface_tension.db*, and a second data source, the Microsoft® Excel®-file *Polysorbate.xlsx*, with various filtering options (see example file *surface_tension.pbix* available on https://github.com/mj-hofmann/Data-Management-for-Natural-Scientists) on each of the levels *data*, *information* and *parameters*. In Microsoft® Power BI Desktop®, data originating from various types of sources can be combined and jointly visualized.

set of *visuals*. Despite its main focus on business intelligence, Microsoft® Power BI Desktop® also has its advantages for visualizing natural scientific data. Particularly useful visuals are

- line plot,
- scatter plot,
- bar plot,
- slicer and
- table.

These types of visuals are also used in Figure 7.15. The general idea of a visual is to select the type of representation you have in mind for a particular set or part of your data, e. g., a surface tension isotherm. To realize this in Microsoft® Power BI Desktop®, we select the visual *line plot* from the list of available visuals. This will introduce some type of blank form on the plotting area on the left-hand side of the interface. Next, we declare that we want to have the surface tension values of the `data` table on the y-axis and the concentration of the `data` table on the x-axis. This is done via drag and drop. How easy is that? At this stage, you end up with a rather shaky zigzag-curve (this intermediate step is not shown in Figure 7.15). This is where the original *business* focus of Microsoft® Power BI Desktop® shines through: If not further specified, it will calculate the *sum* values as default option. To arrive where we want to go, we have to disentangle this cumulative curve by setting the `Sample` column as "Legend" (again via drag and drop). Similarly, we could have used the `CAS` or `id` columns. The procedure is similar for the other available visuals: Make your mind up on what you want to show, drag and drop the relevant columns and do some adjustments for preferences via the rich formatting options—if you are familiar with Microsoft® Excel®, it's a plus. And you are done.

The process of setting up this basic visualization from a given *data model* is not described in full detail here. However, this should serve as a starting point to further search online.

With our visuals ready and shiny, let's move on to the actual benefit of Microsoft® Power BI Desktop® for use in natural sciences: *slicing* and *refreshing*.

Even though available in the list of visuals, slicers do not fit that category perfectly—at least, according to my humble opinion. Yes, you can drag and drop one or more columns into the field of a slicer, but it does not really serve the purpose of visualizing. As the name indicates, it rather cuts down the visualized data according to the specified dimension. A typical use case is to filter for *data* connected to certain values of characteristic *parameters*, i. e., surface tension isotherms corresponding to a certain CMC-threshold. That is a pretty SQL-like query without even a single line of coding but just some drag and drop actions. The beauty of Microsoft® Power BI Desktop® is in its ability to answer these types of questions without even needing to write a single line of code (in the present example) and to provide the ability to seamlessly navigate between the layers of *data*, *information* and *parameters* via a Graphical User Interface (GUI) readily adjustable to specific needs of the user.

The second benefit introduced by Microsoft® Power BI Desktop® is *refreshing*. Assume you collected the experimental raw data of ten more samples, ran the previous Python scripts once more to extract *data* and *information*, obtained the characteristic *parameters* and stored all of them to the SQLite-database *surface_tension.db*. Now you want to review your results.[5] Just hit the "Refresh"-button at the top of Microsoft® Power BI Desktop®'s interface and all your results will be available for reviewing. You are interested in just the latest samples? Simply set an appropriate slicer filtering, e. g., for date of measurement or date of analysis, to boil down your data according to your needs. Other than in woodworking, where it's *measure twice, cut once*, in performing experiments and visualizing them via Microsoft® Power BI Desktop®, it's *measure once, slice as many times as you want.*

---

5 Of course, looking at raw data should not be done only after the *parameters* have been extracted, but rather early in the overall process to spot and counteract potential errors during the measurement or processing steps.

7.4 Wrap up

This chapter introduced multiple options for sourcing data (file structure and SQLite-database), selecting relevant parts for the specific questions to be addressed (via `pandas` and `SQLAlchemy`) and plotting (relying on `matplotlib` and/or `seaborn`). Furthermore, visualization of the experimental *data*, *information* and *parameters* via the business intelligence solution Microsoft® Power BI Desktop® was demonstrated. The essential steps for connecting to data sources and building a *data model* (how are the tables related?) were presented.

The chapter concludes with an exemplary summary visualization of the *combined* contents of the SQLite-database *surface_tension.db* and the additionally available Microsoft® Excel®-file *Polysorbate.xlsx*. Connecting the latter data sources enables filtering across the initial data source boundaries and the levels of *data*, *information* and *parameters*.

8 Responding to lessons learned

> *The foolish and the dead alone never change their opinion.*
> James Russell Lowell, 1871

In most real-world cases, there's some knowledge generated in the course of planning, conducting, analysing, pondering and communicating about your experiments. Depending on the scope of your project, the overall process may take weeks, months or even years. It is purely natural that you did not and could not have thought of every aspect of possibly relevant ways of looking at your experiments from the beginning. If you knew the results or outcome right from the start, what would (actually) be the point in carrying out the experiments in the first place?

Anyway, the point I am trying to make here is to make you aware of the powerful opportunity to look at your data from a different perspective even if *the* analysis, i. e., the one you originally had in mind, has been done. Keep going and digging further. Gaining insight is typically a highly iterative process.

8.1 Taking another perspective

As indicated in Figure 8.1, having an explicit view of a research subject from a certain perspective is not at all wrong. You might, however, make a mistake in rejecting someone else's equally strong view if it does not match yours. As this admittedly suggestive sketch points out, you tend to get a more general, less personally biased view of things the more perspectives you take into consideration. This tends to be a lot easier—and also more fun—if you can do this with other people around you. In the best of all cases, this should occur in a positive, target-oriented and respectful joint effort. You do need to be friends, but the relationship should be sufficiently functionally sound to enable a free exchange of ideas and challenging each others' hypotheses.

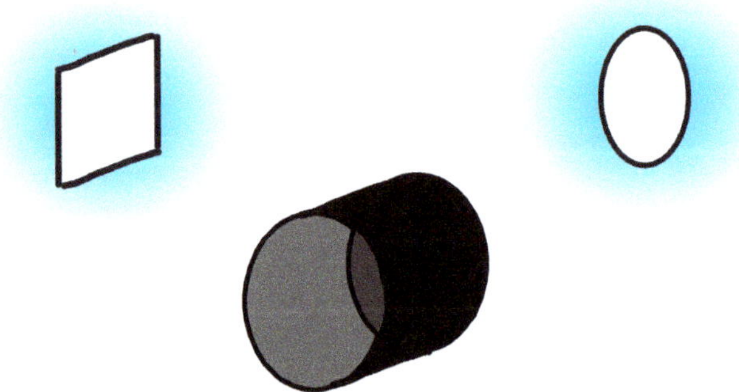

Figure 8.1: Different perspectives on "one" (research) subject.

https://doi.org/10.1515/9783110788433-008

Now, returning from people to data: Ideally and for the sake of the conciseness of this book, we assume to have collected the *right* experimental raw data. This means that there's no need to subject the samples to any further measurements and that we are able to extract some additional *parameters* from

- the collected *data* alone, such as maxima, minima, peak positions, onsets etc. or
- *data* and collected *information* such as the equilibration behaviour of a measured property given certain conditions (such as temperature or humidity) or
- *data* and so-far missing *information*.

The latter case is obviously the most challenging situation. A typical example of this is missing temperature and/or humidity information corresponding to times of sample preparation and measurement. If you are lucky, your labs are equipped with a logged thermostat that collects temperature and humidity in regular intervals. If you know when and where your samples were stored, prepared and measured, you are good to go: Map the thermostat data to the relevant times corresponding to your samples. Additionally, you should consider documenting temperature and humidity information in the next set of experiments of this kind immediately.

8.2 Extracting and ingesting additional parameters from data

In the next pages, we will cover the scenario in which a certain *parameter* obtainable from our *data* should have been collected but was not—for whatever reason. This insight can potentially arise from the analysis conducted so far, exchanges with colleagues or divine revelation: Anything is welcome. Here, we will extract the *minimum obtainable surface tension* for each of the studied samples. Just as the critical micellar concentration (CMC), it is a relevant quantity in certain applications. Often, the purpose of a surfactant is to reach a potentially low surface tension at minimum concentrations, i. e., a low CMC to minimize the surface active agent's use, and thereby the related cost and risk to the environment.

Having access to the *data*, extraction of further *parameters* is still possible in hindsight. Herein, a way to add additional *parameters* to an already existing set of *data* and *information* is shown using an SQLite-database via Python and `SQLAlchemy`. One last time, pseudo-code for the following steps is given in Listing 8.1.

Listing 8.1: add_new_parameter_pseudocode.py

```
# %% 1) read relevant existing SQLite-database tables
#

# %% 2) extract "new parameter"
#

```

```
# %% 3) write modified "parameter" table back to database
#

# %% 4) visualize new findings
#
```

For convenience, we will use the SQLite-database as our single data source and extract the `data` and `parameters` tables from it. Extraction of the *new parameter* is done via `pandas`, and the resulting values will be appended to the `parameters` table. In order to maintain the intended relationship between these tables, this step is carried out via `SQLAlchemy`. A check for success is achieved through a final plotting of the adapted table contents. To begin with, we have to establish a connection to our SQLite-database as in the previous chapter. Once again, this is done explicitly in Listing 8.2 because we did not define a function returning the `engine` and `metadata`.

If you plan to use this "connecting to database"-part on several occasions, defining an appropriate function returning the `engine` and `metadata` objects might be an useful option for you.

Listing 8.2: add_new_parameter.py

```
# %% 1) read relevant existing SQLite-database tables
#

# module for handling SQLite
import sqlalchemy
# for DataFrame handling
import pandas as pd

# SQLite filename
file_sqlite = "surface_tension.db"

# generate connection to database
engine = sqlalchemy.create_engine(
            "sqlite:///" + file_sqlite,
            echo=True
            )

# create bound metadata
metadata = sqlalchemy.MetaData(engine)
```

Loading all available table definitions from the database is done via the `reflect`-method of the `MetaData`-object `metadata`. Further details on both tables and columns can be obtained by an `Inspector`-object. In Listing 8.3, we use the `inspector` to recall our table names and the columns already contained in the parameters table via the inspector's methods get_table_names and get_columns, respectively.

Listing 8.3: add_new_parameter.py (continued)

```
# load all available table definitions from the database.
metadata.reflect()

# define "inspector"
inspector = sqlalchemy.inspect(engine)

# show avaiable table_names
for _table_name in inspector.get_table_names():
    # show this table
    print("   -", _table_name)

#   - data
#   - information
#   - parameters

# show columns in "parameters" table
for _column in inspector.get_columns("parameters"):
    # print column name
    print("   #", _column["name"])

#   # cmc_g_l
#   # id
```

Next, we will get the tables from our SQLite-database. It is to be noted, however, that we can access those on two different levels. The corresponding code is shown in Listing 8.4.

Using the pd.read_sql_table-function with the table name and the connection as arguments returns a DataFrame. This can be used for our conventional selection, transformation and other tasks in the Python world.

Access via the metadata.tables returns a sqlalchemy.sql.schema.Table. This object provides only information on the structure, relationships and constraints of the table with respect to other tables within the database. Here, no content is involved.

Listing 8.4: add_new_parameter.py (continued)

```
# get "data" as pd.DataFrame
data = pd.read_sql_table("data", engine)
# get "parameters" as pd.DataFrame
parameters = pd.read_sql_table("parameters", engine)

# get "parameters" and "data" as sqlalchemy.Tables
```

```
table_parameters = metadata.tables["parameters"]
table_data = metadata.tables["data"]

# return types depending on kind of "table access"
print("data is of type", type(data))
# data is of type <class 'pandas.core.frame.DataFrame'>
print("table_data is of type", type(table_data))
# table_data is of type <class 'sqlalchemy.sql.schema.Table'>
```

Having our *data* at hand, we can now move on to the actual task: extraction of our *new parameter* from the collected experimental data. To get a visual representation of what we are going to look for in each sample, see Figure 8.2. We are interested in the minimum attainable value of surface tension *larger than* the respective CMC. For simplicity, we assume (and also graphically verify this assumption) that the target value equals the readily available minimum value of surface tension in the entire sub-dataset corresponding to an individual sample according to Figure 8.2.

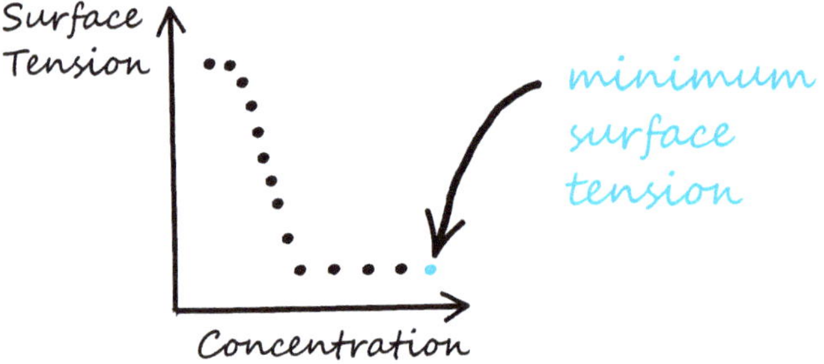

Figure 8.2: Schematic representation of the extraction of the new parameter *minimum surface tension*.

Once more, we will loop over our entire dataset, i. e., perform the analysis for each sample using the groupby-method. Within this process, we fill an initially empty `dict` with the appropriate key-value pairs, whereas the key is the sample name and the value is our newly extracted parameter. Other than in Listing 8.5, the characteristic parameter extraction task can be, and most probably is, more intricate than in this basic showcase. Nonetheless, the concept remains unchanged also for more complex situations.

Listing 8.5: add_new_parameter.py (continued)

```
# %% 2) extract "new parameter"
#

# plotting library for visual check
```

```
import matplotlib.pyplot as plt

# initialize dict holding the new parameter "minimum surface tension"
parameter_to_add = dict()

# loop samples
for _g, _data in data.groupby(by="id"):
    # info
    print(_g)

    # get minimum surface tension value
    min_surf_tension_mN_m = _data[
        "surface_tension_mN_m"
        ].min()

    # info
    print(min_surf_tension_mN_m)

    # add value to dict
    parameter_to_add[_g] = min_surf_tension_mN_m

    # plot for plausibility check
    # all measured points
    plt.plot(
        _data["concentration_g_l"],
        _data["surface_tension_mN_m"],
        marker="o"
        )
    # horizontal (--> h) line to highlight identified
    # minimum surface tension
    plt.axhline(
        min_surf_tension_mN_m,  # y-value of horizontal line
        color="red",
        alpha=0.5  # opacity
        )
    # set to log scale
    plt.xscale("log")
    # add title
    plt.title(_g)

    # plot cosmetics
    plt.xlabel("Concentration / [g/l]", size=14)
    plt.ylabel("Surface tension / [mN/m]", size=14)

    # save (last) plot
    plt.savefig(
        "minimum_surface_tension_example.png",
        dpi=300
        )
```

```

    # show plot
    plt.show()
```

As an example for visualizing the minimum surface tension obtained for the *Polysorbate85.csv*-sample, see Figure 8.3.

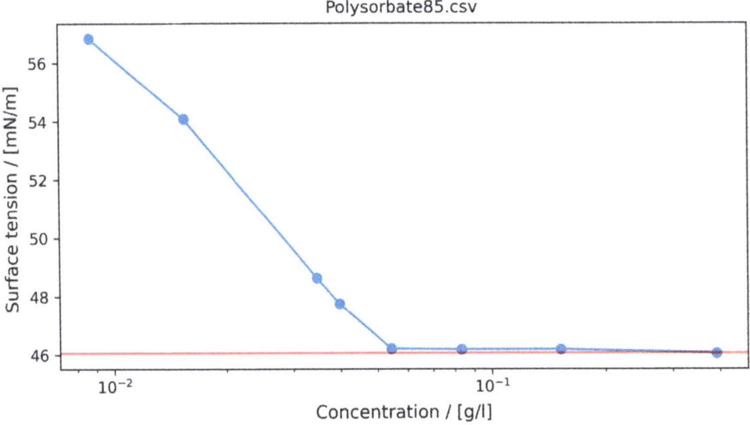

Figure 8.3: The new parameter *minimum surface tension* collected for the *Polysorbate85.csv*-sample indicated by a horizontal line.

8.3 Extending the existing database table `table_parameters`

Next, we have to introduce these newly obtained parameter values into our already existing database table `parameters`. A schematic representation is given in Figure 8.4. Taking a closer look, two successive steps have to be carried out. First, we have to create some space for the new parameter, i. e., to append a new column to the existing table leaving everything else unchanged, in particular the relationships and constraints. Second, we need to "sort in" our new data appropriately, i. e., we need to ensure that our new parameter values are mapped appropriately to the existing entries. For that reason, we used a `dict` rather than a `list` for storing our freshly baked parameters.

In order to ingest the desired changes to our SQLite-database, we use an extension to the `SQLAlchemy` library: `sqlalchemy-migrate`.[1] This package extends `SQLAlchemy` to have *database changeset handling*. An alternative migration tool is `alembic`.[2] With this module imported as shown in Listing 8.6, we can use the `create_column`-method of

1 https://pypi.org/project/sqlalchemy-migrate/

2 https://pypi.org/project/alembic/

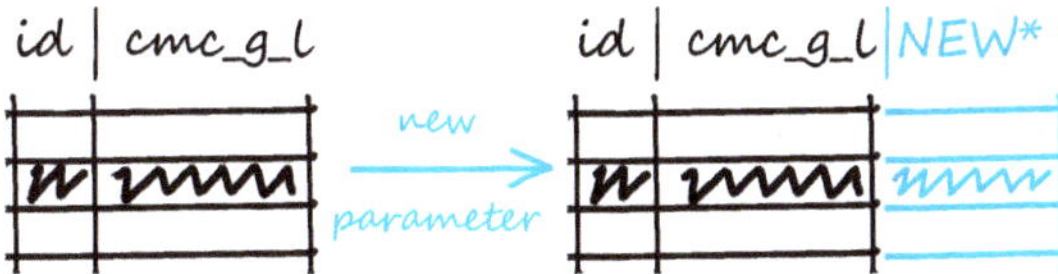

Figure 8.4: The extracted parameters corresponding to individual samples are stored in an additional column of the already existing table `parameters`. Within the process, the required space is created as a first step.

our table `table_parameters` to define (or append) a new column. Therein, the column to be appended can be defined with constraints, primary and foreign key conditions. Note, that these changes are *immediately effective* in the database. Running the code a second time will trigger an `OperationalError` due to an intended duplication of column name.

Listing 8.6: add_new_parameter.py (continued)

```
# %% 3) write modified "parameter" table back to database
#

# module for table modifiaction
import migrate.changeset

# "append" new columns to existing Table
table_parameters.create_column(
    sqlalchemy.Column(
        'minimum_surface_tension_mN_m',
        sqlalchemy.Float
        )
    )
```

If you wish to run the code multiple times, find a remedy for the previously mentioned `OperationalError` via exception handling: Place the column creation code in a `try-except` block.

8.4 Writing to the extended table table_parameters

Once the appropriate space for storing the new parameter values has been created, writing the actual values to the database is the only thing left to do. As the entries are already partly existent, we just have to UPDATE an existing entry, as it is coined in the Structured Query Language (SQL)-world. In this context, an entry is a row in our database. Immediately after the creation of the new column, it is empty. Our updating task is to write exactly this missing value to the database in the appropriate place. This means we write the value of our `parameter_to_add` `dict` to the newly defined `minimum_surface_tension_mN_m` column of our table, where the `dict`s key equals the entry in the table's `id` column. This is stated more clearly and concisely in Listing 8.7:

Listing 8.7: add_new_parameter.py (continued)

```
# update values
for key, value in parameter_to_add.items():
    #info
    print("Inserting", value, "for", key)
    # update table
    table_parameters.update().where(
            table_parameters.c.id == key  # at specific row
            ).execute(
                minimum_surface_tension_mN_m=value  # set the value
            )
```

Again, we will conclude our new parameter ingestion task by visualizing some *data* and *parameters*. To make use of the defined relationship between the tables `table_data` and `table_parameters`, we apply the first table's `join` method according to Listing 8.8. Note that the resulting variable is of type `Join`.

Listing 8.8: add_new_parameter.py (continued)

```
# join "data" and "parameters" tables according to declared keys
join = table_data.join(table_parameters)
```

For defining our variable `query` of type `Select`, we use the SELECT-function of `SQL-Alchemy`, specify the columns we would like to obtain and take into account the previously defined `join` as the argument of the `select_from`-method. In this example, we use all columns from `table_data` and the columns `minimum_surface_tension_mN_m` and `cmc_g_l` from `table_parameters`. As shown in Listing 8.9, a selection of columns is possible via the list within the SELECT-statement. Finally, the query is once more materialized by the `read_sql`-function of `pandas` using the defined `query` and the connection via the `engine` variable as arguments.

Listing 8.9: add_new_parameter.py (continued)

```
# select
query = sqlalchemy.sql.select([
            table_data,  # use all columns from "data" table
            table_parameters.c.minimum_surface_tension_mN_m,
            table_parameters.c.cmc_g_l  # select CMC column
```

```
                ]).select_from(
                    join  # used joined tables as "pool" to select from
                    )

# get DataFrame corresponding to query
query_data = pd.read_sql(query, engine)
```

The columns available within an object of type sqlalchemy.sql.schema.Table are accessible both via the c and columns' attributes. Try using either variant or a combination thereof. In the previous code snippet, table_parameters and table_data are variables of this type.

In order to visualize the results both on the *data* and the *parameter* level, we use the seaborn library in combination with matplotlib. Here, we want to create an overall figure with two subplots: a scatter plot of the newly introduced parameter *minimum surface tension* against the CMC and the corresponding surface tension isotherms. According to Listing 8.10, both plots are colour-coded via the id-column used as the value for the plotting-function's hue-key.

Listing 8.10: add_new_parameter.py (continued)

```
import matplotlib.pyplot as plt
import seaborn as sns

# define 2x1 plot setup
fig, (ax1, ax2) = plt.subplots(1, 2)

# define colour palette to be used
# see https://seaborn.pydata.org/tutorial/color_palettes.html
palette = "Paired"

# scatterplot of minimum surface tension vs CMC
sns.scatterplot(
    data=query_data,
    x="cmc_g_l",
    y="minimum_surface_tension_mN_m",
    hue="id",  # set colour by id
    palette=palette,
    ax=ax1,
    legend=None
    )

# line plot of correponding surface tension isotherms
sns.lineplot(
    data=query_data,
    x="concentration_g_l",
    y="surface_tension_mN_m",
```

```
    hue="id", # set colour by id
    palette=palette,
    marker="o",
    ax=ax2
    )

# set logarithmic axes
ax2.set_xscale("log")

# set label explicitly
ax1.set_xlabel("CMC / [g/l]")
ax1.set_ylabel("Minimum surface tension / [mN/m]")
ax2.set_xlabel("Concentration / [g/l]")
ax2.set_ylabel("Surface tension / [mN/m]")

# legend finetuning
plt.legend(
    frameon=False,  # no frame
    fontsize=8  # smaller font size
    )

# save plot
plt.savefig(
    "add_new_parameter.png",
    dpi=300
    )
```

The resulting plot is saved as a *png*-file and shown in Figure 8.5.

8.5 Wrap up

This chapter introduced the notion and necessity of understanding *data processing and analysis* as iterative procedures. Insights concerning alternative perspectives of looking at the so-far collected *data*, *information* and *parameters* should be appreciated. For this purpose, the extraction of additional *parameters* from already existing *data* was demonstrated. Extending an available SQLite-database table via `SQLAlchemy` was shown as a next step comprising both the allocation of a new column and writing the freshly extracted parameter `minimum_surface_tension_mN_m` to it.

The chapter concludes with visualizing parts of the modified SQLite-database *surface_tension.db* for the purpose of validating the induced changes and utilizing the newly introduced parameter.

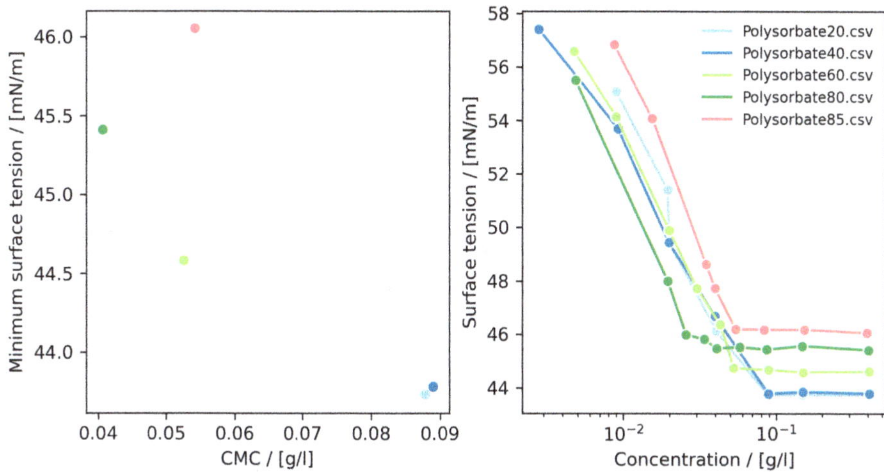

Figure 8.5: Left: `scatterplot` from `seaborn` of minimum surface tension against CMC (*parameter* level). Right: Corresponding surface tension isotherms obtained as `lineplot` from `seaborn` (*data* level).

9 Where to go from here

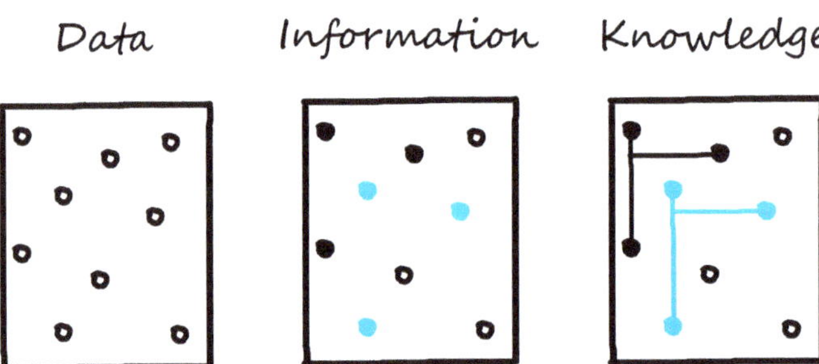

So are you still aiming to reach the stages of *knowledge* or even *wisdom* according to the previously introduced *data pyramid*? With our *data* and *information* structured, we ensure that the data is FAIR [7]. The FAIR guiding principles for data management and data stewardship aim to provide guidelines to improve the findability, accessibility, interoperability and reuse of digital assets.[1] Meeting these guidelines is beneficial for both human operators and processing via machines in particular. The latter preferably rely on a well-defined structure (with the exception of complex machine learning approaches) within data sets to reach peak performance. The FAIR-principles in essence are:

Findable: First things first, we need to find the data to use and reuse them. When it is easy for you as a human to find data and metadata, this is probably also the case for a computer.

Accessible: After finding it, you need to access the data. This might include authentication and authorisation depending on the policy of your organization.

Interoperable: Typically, there is a need for integration with other data. In addition, the data needs to interoperate with applications or workflows for analysis, storage and processing.

Reusable: the supreme discipline and ultimate goal of data management. Adequate combination of *data* and *information* ideally allows for addressing various questions from the data at hand if they are queryable according to various aspects adopted to the respective use case.

With the FAIR-guidelines in mind, there is at least the chance to move towards the upper levels of the data pyramid: *knowledge* and *wisdom*: There are multiple types of analyses, each related to techniques popularized by the huge field of data science. In order of increasing complexity, those are

1 https://www.go-fair.org/fair-principles/

https://doi.org/10.1515/9783110788433-009

- descriptive,
- explorative,
- inferential,
- predictive and
- causal analysis.

In the context of natural sciences, we are most typically involved and interested in the latter two cases or stages of analysis: predictive and causal. We want to know how a system behaves if we turn the metaphorical "knob" (predictive analysis) and, in the best of all worlds, we get time (and funding) to find out *why* this behaviour occurs (causal analysis).

9.1 Linear models

A first step towards modelling of relationships between variables is linear modelling. Despite their simplicity, linear models are frequently used. It is even their simplicity which allows deducing a clear and readily understandable cause–effect pattern, e. g., if the concentration of substance A is increased, the reaction yield increases by a certain amount assuming all other things being equal. If we take a closer look at the plot in the centre of the bottom row in Figure 7.15, we can observe a relation of this kind between the critical micellar concentration (CMC) and molar mass of the surfactant. As this visualization suggests, a higher molar mass comes along with a lower CMC. This also makes sense from a physico–chemical perspective. The structurally related surfactant molecules described in the present minimalistic database show an increased tendency towards aggregation with higher molar mass. This corresponds to a lower CMC-value. In other words: At a given concentration, higher molecular-mass surfactants are more susceptible to aggregation. Please note that in this scenario, the *all other things being equal* condition refers also to the structural type of the surfactant. Of course, there are—in the large world of detergents—lower molecular weight molecules having much lower CMC values than the presented polysorbates.

At this stage, it is essential to not blindly rely on the results of the model, but rather to take into account your domain knowledge. Probably, you can come up with a more complex model taking into account not only molecular mass but also other distinguishing features of the surfactants. There are essentially no limits except your imagination—and the reality of experimentally gained results. Maybe, introducing one or two additional quantifiable *parameters* to your model allows for a better description of the experimental results with your model. With this idea in mind, we moved from linear regression to its more general analogue, *multivariate linear regression*. The idea is to predict a value based on two or more variables. The Python package `scikit-learn` helps solving questions of this type.[2] I will not go into details

2 https://scikit-learn.org

concerning this library because there are many helpful resources available via a quick online search.

To sum up, the choice of model complexity is also critical for models as simple as the described linear models. The well-known *bias–variance trade-off* describes the conflict between *underfitting* and *overfitting*. I would like to introduce an example to highlight the key ideas: Let's assume you are challenged to predict the height of a random adult person in your circle of friends. What is your answer?

Maximum biased answer: If you do not have any further information, your best guess would be the average value, say 1.75 m.

Less bias, more variance: You are given the further information that the person is female. With this you can extend the model to get two different predictions, an average value for your male friends and one for your female friends. In our example, this would lead to the prediction of a height of 1.67 m, let's say.

Less bias and even more variance: You are further given the information, that the person's weight is 55 kg. Taking into account this information allows for obtaining a prediction from a model using gender and weight as variables. In our example, this will lead to a lower body-size prediction to 1.64 m, let's say.

No bias: You get the information that the random person is Julia. You know that her height is 1.70 m, so that is the value you would predict.

Clearly, this is no longer a "prediction" but just *reading an individual value from a list*. In practice, determining where to stop the addition of ever more *parameters* is a challenging task and requires both domain and statistical knowledge.

There are, however, great software tools that can help in the construction of multivariate linear regression models, including, but not exclusively:
- MODDE®,[3]
- MATLAB® and[4]
- the just-mentioned `scikit-learn` library of Python.

Next to linear regression models, there is a large and growing number of additional complex approaches, such as *neural networks* or other *machine learning* techniques applicable for the analysis of—preferably—larger data sets. As this book aims to serve merely as a starting point for further activities, I will leave this to the dedicated literature [4].

9.2 Causal analysis

A widely neglected pathway of analysis in the field of natural sciences is probably *formalized causal analysis*, where the aim is to draw cause-and-effect relationships

---

3 https://www.sartorius.com/en/products/process-analytical-technology/data-analytics-software/support/knowledge-base/modde-regression-coefficients-used-equation-excel-550614

4 https://www.mathworks.com/help/stats/linearmodel.html

between observable and tunable variables and outcomes. This point touches the very essence of the scientific approach, so let me give you an example.

In general, natural scientists are in a very comfortable situation compared to their colleagues from, e. g., the social sciences. If interested in a certain *cause-and-effect relationship*, we set up an experiment, modify the assumed causation and measure the effect on a particular response variable. In a simple experimental setup, we might be interested in the influence of temperature on viscosity for a particular pure liquid as sketched in Figure 9.1.

Figure 9.1: Schematic influence of temperature on viscosity or more generally: *A causes B*.

Let's move one step further. Assume we use this liquid as a catalyst in a well-defined model system. In order to determine the overall system performance, we experimentally capture a performance parameter to quantify the outcome. Variables suspected of influencing this parameter are the experimental temperature, pH-value and viscosity. In a first basic model, let's further suppose that experimental temperature influences directly both the experimentally determined performance value and viscosity (just as in the previous basic example). Furthermore, we will assume that viscosity also has a certain influence on the observed performance value. For the chemists and material scientists among you, this modified performance might be due to increased or lowered mobility of reacting molecules accompanying changes in viscosity. Accordingly, the sketch shown in Figure 9.2 serves as a graphical model of our hypothesis to be tested. Therein, an arrow from one property to another reads as *influences*, e. g., temperature influences viscosity.

9.3 dowhy?

With this basic considerations, we set the stage for systematic causal analysis. Herein, I'll show this approach via the dowhy-package (read as *do why?*) [5].[5] It aims to spark causal thinking and analysis via a readily accessible *four-step interface* directed towards nonexperts in the field of machine learning, i. e., the well-known *practitioner*. The site referred to in the footnote offers a lot more information in greater detail, which will not be covered here. The four steps for conducting a causal inference analysis are:

5 https://microsoft.github.io/dowhy

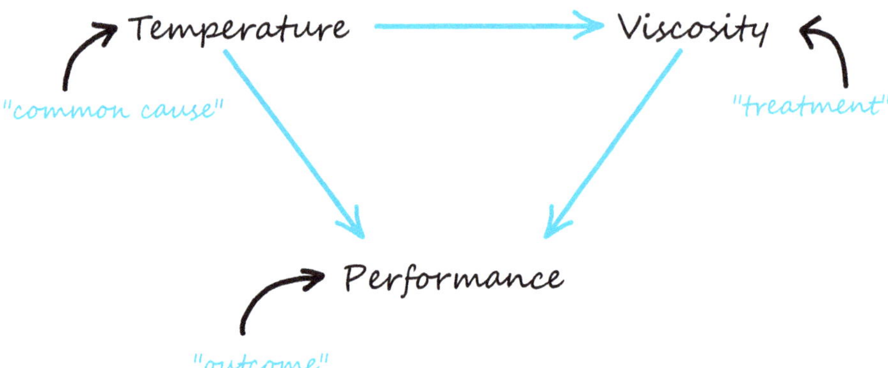

Figure 9.2: Causal graph representing a hypothesis on the relationship between experimental parameters and results. In terms of causal analysis, *Performance* is considered as the *outcome*, *Viscosity* as the *treatment* and *Temperature* as *a common cause* influencing both viscosity and performance. The question to be answered in the following is whether viscosity (*treatment*) has a causal effect on the observed values of performance (*outcome*), taking into account all available data.

1. modelling a causal inference problem using assumptions, i. e., developing a hypothesis to be tested,
2. identifying an expression for the causal effect under these assumptions, i. e., the *causal estimand*,
3. estimating the expression using statistical methods, such as matching or instrumental variables and finally
4. verifying the validity of the estimate using a variety of robustness checks.

To summarize: We first build a model based on a hypothesis, which has to be visualized via a *graph* (see Figure 9.2). Then, we use the rigorous approach provided by the dowhy-package to check if our experimental *data* fit the outlined model.

! In general, *data* is *not* fitted to a model, but model parameters are adjusted or "tuned" to agree with or "fit" the experimental results as represented by the collected *data*.

If the model stands the test, it's a good sign. If not, build another hypothesis and run the fitting and checking steps once more. Here, *fit* is rather to be understood as *agrees with*. In the following example, a set of synthetic data is generated via dowhy and analysed via the described four step approach. The data generation process and a basic visualization of the results is given in Listing 9.1. The essential part of this code snippet is the generation of an artificial dataset using the `xy_dataset`-function. In a real world scenario, we would of course use experimentally determined characteristics and observed parameters. Here, we use 20 sample points, which is assumed to be a realistic number in many use cases.

Listing 9.1: causal_analysis_example.py

```
# based on
# https://microsoft.github.io/dowhy/example_notebooks/tutorial-causal
# inference-machinelearning-using-dowhy-econml.html
# https://microsoft.github.io/dowhy/example_notebooks/dowhy_confounder
# _example.html?highlight=slope

# imports
import matplotlib.pyplot as plt
# import two modules from "dowhy" in one line
import dowhy.datasets, dowhy.plotter
import seaborn as sns

# set treatment (viscosity) as causal for the outcome (performance)
effect = True

# build sample data set
data_dict = dowhy.datasets.xy_dataset(
                20,  # number of samples
                effect=effect,
                num_common_causes=1,
                is_linear=False,
                sd_error=0.75
                )

# get DataFrame
df = data_dict['df']
# rename columns to match the envisioned example
df.columns = ["Viscosity", "Performance", "Temperature", "pH"]
# tune values
df["Temperature"] += 20
df["Performance"] += 75
# show info / DataFrame structure
print(df.head())

# specify "outcome", "treatment" and "cause"; these are the terms
# used in causal analysis
outcome = "Performance"
treatment = "Viscosity"
# one cause
common_causes = "Temperature"
# # two causes
# common_causes = ["Temperature", "pH"]

# plot sample data via (dowhy built-in) function
dowhy.plotter.plot_treatment_outcome(
        df[treatment],
        df[outcome],
        df["pH"]
        )
# show plot
plt.show()

```

```
# make custom "plot_treatment_outcome"-like plot
x = "pH"  # use "pH" as variable on the x axis
for _c in [treatment, outcome]:
    plt.scatter(
        df[x],
        df[_c],
        label=_c
        )
# plot cosmetics
plt.xlabel(x)
plt.ylabel("\n".join([treatment, outcome]))
plt.legend(frameon=False, fontsize=12)
# save custom plot
plt.savefig(
        "modified_plotter_type.png",
        dpi=300,
        bbox_inches="tight"
        )
# show
plt.show()

# pairplot to show variable dependecies
sns.pairplot(
        df,
        height=3.5,
        corner=True,  # don't add axes to the upper triangle
        diag_kind="hist"
        )

# save pairplot
plt.savefig(
        "df_pairplot.png",
        dpi=300
        )
```

The remainder of the previous code is dedicated to adjusting the example to a more natural scientific type of a problem and shifting the values of the variables `Temperature` and `Performance` to somewhat more reasonable ranges. In the chosen example, we want to determine the causal effect of viscosity on performance (which is there because we set it to be there). Nevertheless, a visualization of the obtained dataset is a helpful first step for working with the *data*. Therefore, we can use `dowhy.plotter`'s function `plot_treatment_outcome`, which does exactly what its name suggests. This function, however, allows no further customization in terms of label naming, legend entries or further parameters. Therefore, `matplotlib` is used to generate an adapted `plot_treatment_outcome`-like plot as shown in Figure 9.3. Furthermore, we use `seaborn`'s `pairplot`-function to get an overall view of the data as shown in Figure 9.4.

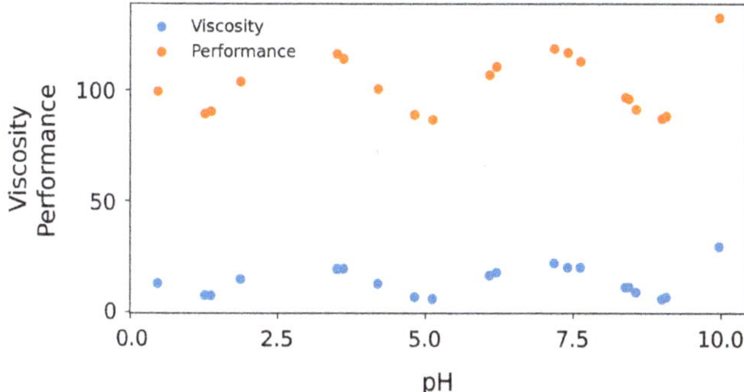

Figure 9.3: Custom `plot_treatment_outcome`-like plot for highlighting the behaviour of the assumed outcome (performance) and treatment (viscosity) against a third parameter (pH) for the generated artificial dataset. This visualization serves to provide an overview of the available dataset.

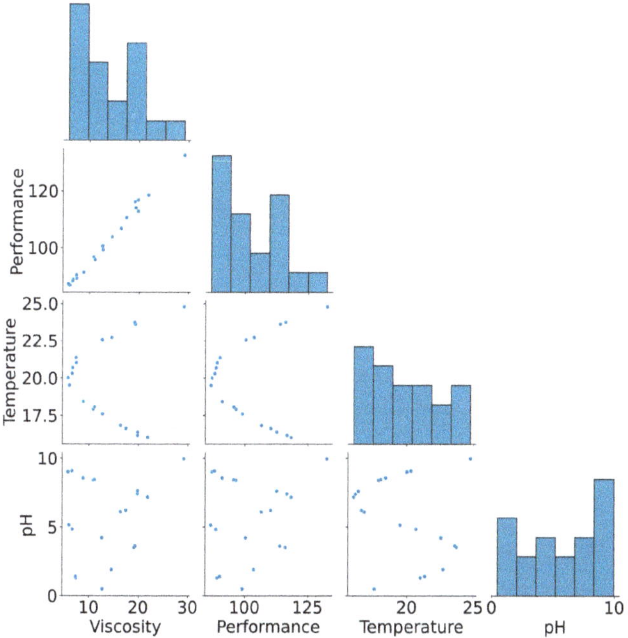

Figure 9.4: The generated dataset visualized via seaborn's `pairplot`-function. In the following, the relationship between `Performance` and `Viscosity` will be evaluated in more detail using the causal analysis approach provided by the dowhy package.

9.3.1 Model the problem as a causal graph

The first step of causal analysis is the specification of a *causal graph*. This is also the place, where expert knowledge is a valuable source of input. If there are thoughts and experienced knowledge on which parameter influences another, we can take this into account right here. It is *the* basic step of the process. The following steps can—on a very high level—be understood as a *validity check*: Do the so-far experimentally obtained values disprove or *refute* the model, i. e., the causal graph?

An important point should be made here: Agreement of your *data* with the assumed model specified via the causal graph does not actively prove the model right. Instead, the result should be understood in a way that the present *data* available do not contradict with the assumed model. The model is "just" not wrong—at the moment, taking into account the so-far considered/available *data*.

With that said, let's move on to building the causal model for our example. We need to specify the DataFrame for which the causal graph will be tested, and declare treatment, outcome and so-called common_causes, i. e., variables having an influence on both treatment and outcome. Certainly, there are further options available for specifying more complex causal graphs for the more ambitious reader. Once specified, the causal graph (see Figure 9.2) can be visualized according to Listing 9.2.

Listing 9.2: causal_analysis_example.py (continued)

```
# %% step 1: Model the problem as a causal graph
#

# define model, i.e. build causal graph
model= dowhy.CausalModel(
          data=df,
          treatment=treatment,
          outcome=outcome,
          common_causes=common_causes
          )

# show model
model.view_model(layout="dot")

from IPython.display import Image, display

# save model
display(Image(filename="causal_model.png"))
```

9.3.2 Identify the causal effect

Next, we move to the formal part of the analysis. The properties of the causal graph allow for an identification of the causal effect within the model. The algorithm implemented in dowhy takes into account connectivity and directionality between parameters introduced via the causal graph. In Listing 9.3, this essential process is achieved within a few lines of code only. The result of this step is the so-called *estimand*, a "plan on how to resolve the causal question to be answered".

Listing 9.3: causal_analysis_example.py (continued)

```
# %% step 2: Identify causal effect using properties of the formal
# causal graph
#

identified_estimand = model.identify_effect(
        proceed_when_unidentifiable=True
        )
print(identified_estimand)
```

9.3.3 Estimate the causal effect

Estimating the causal effect using dowhy is as easy as providing the `estimate_effect` method of the model object with the previously determined estimand and a method for estimating the effect. In our case, we use a linear regression method.

Also for the representation of the causal effect, there is a built-in function for visualization: `dowhy.plotter`'s function `plot_causal_effect`. Again, this basic plotting tool does not take into account the actual variable names but gives a basic visual representation of the results obtained via Listing 9.4 as shown in Figure 9.5.

Listing 9.4: causal_analysis_example.py (continued)

```
# %% step 3: Estimate the causal effect
#

estimate = model.estimate_effect(
           identified_estimand,
           method_name="backdoor.linear_regression"
           )

print(estimate)
# ## Realized estimand
# b: Performance~Viscosity+Temperature
# Target units: ate

# ## Estimate
# Mean value: 1.876112877792039
print(f"DoWhy estimate of causal effect is {estimate.value}")
# DoWhy estimate of causal effect is 1.876112877792039

# Plot slope of line between action and outcome = causal effect
dowhy.plotter.plot_causal_effect(
    estimate,
    df[treatment],
```

```
    df[outcome]
    )
# show
plt.show()
```

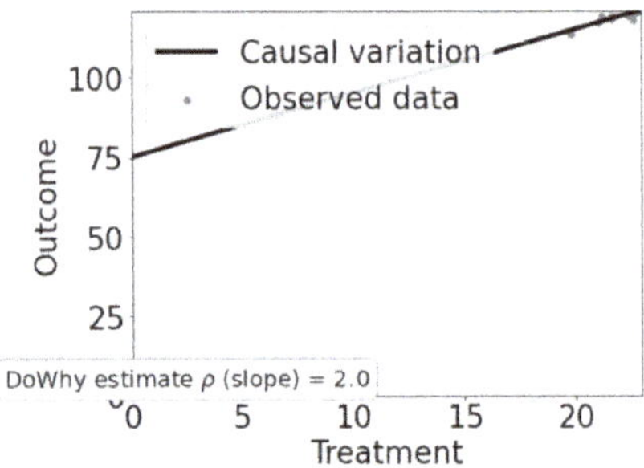

Figure 9.5: Visualization of the causal relationship between performance (outcome on the y-axis) and viscosity (treatment on the x-axis) as directly obtained via the function `plot_causal_effect` of dowhy.

The resulting `estimate` grants access to further parameters and can be used for a customized visualization according to Figure 9.6.

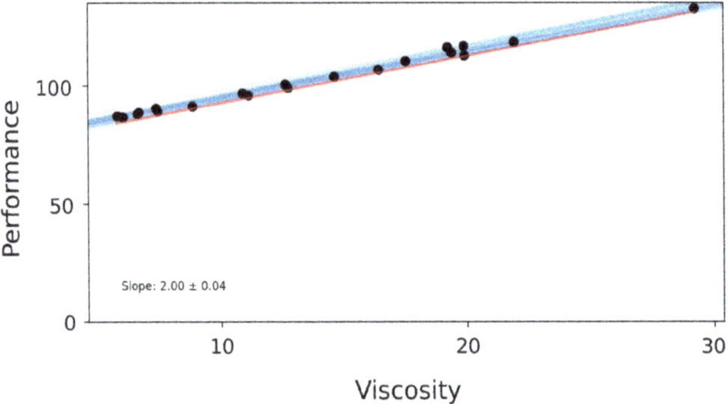

Figure 9.6: Modified causal effect plot obtained via the `estimate` object's attributes. Its `interpret`-method provides a literal description of the obtained results: `Increasing the treatment variable(s) [Viscosity] from 0 to 1 causes an increase of 1.9389466838083251 in the expected value of the outcome [Performance], over the data distribution/population represented by the dataset.` Note that dowhy scales the treatment variable to a range from 0 to 1 as indicated in the previous result message.

To arrive at this figure, the `value` and `intercept` attributes of the `estimate` object were accessed. Furthermore, information on the estimation's errors can be obtained via the corresponding functions as shown in Listing 9.5.

Listing 9.5: causal_analysis_example.py (continued)

```
# %% step 3b: Getting more from the "estimate"
#

# get intercept and slope
intercept = estimate.intercept
slope = estimate.value
# get error information
std_error = estimate.get_standard_error()[0]
ci = estimate.get_confidence_intervals()[0]

# plot "experimental" data and best causal fit
plt.plot(
        df[treatment],
        df[outcome],
        "ko",  # black (k) dot markers (o)
        zorder=1  # plot markers on top layer
        )
plt.plot(
        df[treatment],
        df[treatment]*slope+intercept,
        "r-",  # red solid line
         alpha=0.5,  # opacity
         zorder=1  # plot markers on top layer
         )
# plot fit lines from each experimental point
for _i, _row in df[["Viscosity", "Performance"]].iterrows():
    # info
    for _s in ci:
        plt.axline(
            (_row["Viscosity"], _row["Performance"]),
            slope=_s,
            alpha=.10,
            zorder=0
            )

# get axes
ax = plt.gca()

# info text
plt.text(0.05, 0.1,  # x and y-coordinates
         f"Slope: {slope:.2f} $\pm$ {std_error:.2f}",  # 2-decimals
         transform=ax.transAxes  # use axes coordinate system
         )

# specify y-range
plt.ylim(bottom=0)
# label cosmetics
```

```
plt.xlabel(treatment)
plt.ylabel(outcome)

# get literal results description
estimate.interpret()

# save plot
plt.savefig(
    "modifed_causal_effect.png",
    dpi=300,
    bbox_inches="tight"
    )

# show plot
plt.show()
```

9.3.4 Refuting the estimate

The final step in the causal analysis process is *refutation*, i. e., challenging the results of the estimate. Also for this stage, dowhy provides a convenient interface and multiple options for refutation. In the following, we will use the methods of random_common_cause and placebo_treatment_refuter, but there are many more available in dowhy. All refutation options are aimed at modification of the provided data and observation of whether the estimated effect changes.

In the case of *random common cause refutation*, we observe whether the estimate changes after adding an independent random variable as common cause affecting both treatment and outcome. If the model is robust, the estimation of the causal effect will *not* be altered by the addition of a random common cause variable.

In the *placebo treatment refutation*, the causal effect of treatment on the outcome should go to zero. In this scenario, the true treatment variable is replaced with an independent random variable.

The basic refutation process in our example is shown in Listing 9.6.

Listing 9.6: causal_analysis_example.py (continued)

```
# %% step 4: Refuting the estimate
#

# A) Adding a random common cause variable
res_random = model.refute_estimate(
                identified_estimand,
                estimate,
                method_name="random_common_cause"
                )
```

```
print(res_random)
# Refute: Add a random common cause
# Estimated effect:1.876112877792039
# New effect:1.876864110685575
# p value:0.49

# B) Replacing treatment with a random (placebo) variable
res_placebo = model.refute_estimate(
                identified_estimand,
                estimate,
                method_name="placebo_treatment_refuter",
                placebo_type="permute"
                )
print(res_placebo)
# Refute: Use a Placebo Treatment
# Estimated effect:1.876112877792039
# New effect:0.025135938683012428
# p value:0.48
```

Refutation via the chosen methods confirms the causal effect between viscosity and performance in our example: Addition of a random common cause does not lead to changes in the causal effect, but using a placebo treatment instead renders the causal effect close to zero.

9.4 Wrap up

This chapter introduced the key ideas for advancing from *data* and *information* to the higher levels of the data pyramid: *knowledge* and *wisdom*. The principles of predictive analysis relying on linear models were shown next to the basics of *formalized causal analysis* relying on the dowhy package.

The chapter concludes with a walk through the four-step interface provided by the latter package coined to a natural scientific example.

10 Conclusion

At the end of this book, I would like to reiterate the steps involved in making use of experimental *data* as described herein chapter by chapter and highlight some of the challenges accompanying each of them.

From experimental files to data

Assuming that specific devices dedicated to performing highly specialized measurement tasks are the primary source of *data* in the context of natural sciences, it is of utmost importance to be able to deal with these types of results in an efficient manner. First of all, this means being able to read or parse the respective types of files and disentangle the *data* from the further accompanying *information*. What is the *data*, i. e., the part of the file's content holding scientific information? What is the *information* part (also referred to as metadata) associated with the file? In general, metadata provides information on other data, i. e., in our case on our scientific data. Typical examples of metadata are time stamps, machine and operator information. They are helpful for filtering a larger amount of files, surely. But they do not provide us with any useful hints concerning the scientific question at hand. Getting the *data* versus *information*/metadata distinction right also allows for—*data*-wise—transitioning from one machine to another newer, more advanced and performant one. Typically, machines of this kind will be replaced only once there is no more support provided by the manufacturer or in case of any major defects. Let's assume you are forced—or willing—to invest in new equipment. In this situation, it is pretty certain that the output formats of the devices will not match entirely. You will not be able to compare results from the previous and the new device (precision, accuracy and further scientific aspects set aside for the moment) right away, except you are aware of the process described in Chapter 4. Measuring a certain property using either of the machines, you will be provided with a *data* part relevant for your scientific questions and another *information* part for the purpose of filtering tasks based on metadata.

From data to information

Extraction of meaningful *information* other than the previously discussed metadata from experimental *data* is the playground for experts and domain knowledge. One of the key ideas is to have the *data* part at hand in order to extract this very *information* according to a certain method. In the natural scientific community, this *information* is frequently referred to as *parameter*. No matter how you name it, it concerns a characteristic value representing a larger amount of experimental *data* on the *information* level of the data pyramid.

It is the result of a *data* reduction. One or multiple series of *data* are boiled down to a single number. Inescapably, this comes along with some *loss in translation* proceeding from the *data* level to the *information* level of the data pyramid. Therefore,

https://doi.org/10.1515/9783110788433-010

being able to seamlessly navigate between those levels is highly relevant for sudden plausibility checks. Furthermore, there are many possible ways of looking at the very same set of *data*. Depending on your training, background and experience, you might be interested in very different *parameters* than a colleague. With the *data* at hand, Chapter 5 provides an example of how to derive parameters, i. e., *information* from it.

Where to put data and information

Depending on the scope of your project, the goal and capabilities, there are many options for storing your data in the *right* way. Most easily, this is done via the use of an organized file structure approach that can be realized independently of the operating system. Just create folders or directories holding a specific type of *data*. In practice, this could be one folder for each type of measurement. Preferably you will not have to move the files manually to the target folder, but this will be done automatically upon saving the result from the experimental machine. This simplistic approach naturally comes with some drawbacks. Files may be modified or deleted by anyone having access to the directory with the appropriate permissions—either by mistake or on purpose. Therefore, moving to a database instead of using an ordered file system is a possible solution to this challenge. Modifications by chance are much less likely because it requires some effort to introduce noticeable changes. The issue or risk of potential deletion of multiple or even all entries on purpose by a user with the appropriate permissions remains, however. But, there should not be people of this kind in your institution anyway. Besides, individuals hampering advances in your transition towards a *data*-democratic sharing culture are expected to become a thing of the past. Sharing is caring. If you are working with a hosted database service instead of a local database such as SQLite, there are regular backups to remedy potential loss of data. In Chapter 6, the basic usage of `SQLAlchemy` is introduced. It allows writing Structured Query Language (SQL)-tables from Python. Independent of the actual type of database used in your application, the syntax remains unchanged for SQL and other dialects such as MS SQL, MySQL or PostgreSQL.

How to visualize data and information

We have all been there: hearing colleagues talking about new and exciting results they obtained in the lab just recently. Of course, you are interested in the actual findings and want to take a look. *Seeing is believing*, and therefore visualization of *data* and *information* is probably *the* decisive step in the process outlined in this book for communicating your results. As described in Chapter 7, visualization is where the lovingly curated data comes to life. The steps taken up to this point merely serve as a means to the goal of visualization. Having the *data* and *information* readily available, you "just" need an appropriate idea in mind about which story to tell to convey your message related to your results. Naturally, this is a challenge in its own, and there are many books out there—from helpful to inspiring—on how to craft a compelling story. Herein, visualization options relying on both an ordered folder structure and a SQLite-database are shown. For the actual plotting task, the Python libraries `matplotlib` and `seaborn` were introduced alongside the Graphical User Interface (GUI) based tool Microsoft® Power BI Desktop®. As the graphical representation of (experimental) results is a wide-ranging topic, there is an equally large number of tools varying in complexity and aspiration towards specific aspects. Eventually, there is also a certain degree of personal preference in choosing *your* preferred tool for a specific task. So, try many things and dive deeper into a few selected ones.

Responding to lessons learned

The ability to readily adjust to or "digest" newly obtained learnings is crucial for a development and progress—both on an individual and an institutional level. As exemplified in Chapter 8, the possibility to have a second look on certain experimental *data* from a different point of view, i. e., with another scientific question in mind, is just around the corner. Using `SQLAlchemy` and/or `pandas`, we get access to a specific set of *data*. From this, we can readily extract additional *parameters* and attach these values to an already existing table of characteristics corresponding to specific samples on the *information* level. Depending on the complexity of the analysis, it might be helpful also to store further metadata related to the parameter determination/extraction step. Typical examples are thresholds above or below which a certain operation is carried out or the specification of a reference in case a sample needs to be referenced to a certain standard. With a growing number of characteristic parameters, however, complexity necessarily increases and clarity suffers. It is therefore helpful for a future user of the database (including your future self) to have an unambiguous understanding of the columns' meaning. What does a column represent? What is its unit of measurement? How was it obtained if not directly measured? Are there any more or less meaningful defaults used? A part of these questions can be addressed by a proper naming of the columns, but not every bit of information can be included in the column names for the sake of conciseness. Anything beyond should be included in a *data dictionary*. Data dictionaries are used to provide detailed information about the contents of a dataset or database, such as the names of measured variables, their data types or formats and descriptive texts. A data dictionary provides a concise guide to understanding and using the *data*.[1] Next to the option of manually writing a dictionary, which is useful for the previously described ordered folder structure approach, comments connected to both tables and columns of a database can be specified immediately upon creation using `SQLAlchemy` (see Chapter B).

1 https://data.nal.usda.gov/data-dictionary-purpose

Where to go from here

With both *data* and *information* in hand, we are in a position to reach for the topmost levels of the data pyramid: *knowledge* and *wisdom*. Without diving deeper in the distinction between the latter, a next step in the processing and analysis scheme might be the recognition of patterns via basic approaches such as linear regression models or more intricate machine learning tools to carve out relationships between determined and observed parameters. This may lead in the easiest case to conclusions of the type *When parameter A is low, the parameter C is high*. Based on these observations, experts in the respective fields with a decent domain knowledge will be able to assist in crafting hypotheses to be evaluated and scrutinized. In Chapter 9, a rigorous and formalized approach for determining the *causal* relation between a set of parameters is described by using the dowhy package in a basic example.

As stated on several occasions throughout this book, the overall process is more iterative the further we move away from the experimental raw *data*. Be prepared to not obtain the "right parameters", i. e., *information* right from the start. Be prepared to not determine the right parameters in the "right way" right from the start. And most importantly, be prepared to not have the "right hypothesis" leading the way to *knowledge* and *wisdom*, i. e., a deeper understanding of the findings right from the start. Even if the so-far collected *data* does not contradict your current model and hypothesis: Be prepared to adjust your mental concept of the experimental findings and the relationships between them to the actual results. If conducted properly, experiments don't lie.

A Packaging a custom module

In the following, the creation of a shareable package (to be used, e. g., by colleagues at your institute or the entire community) will be shown. To make use of the `setuptools` package for packaging, a file *setup.py* holding information on the package to be generated is required. The schematic folder structure is given in Figure A.1.

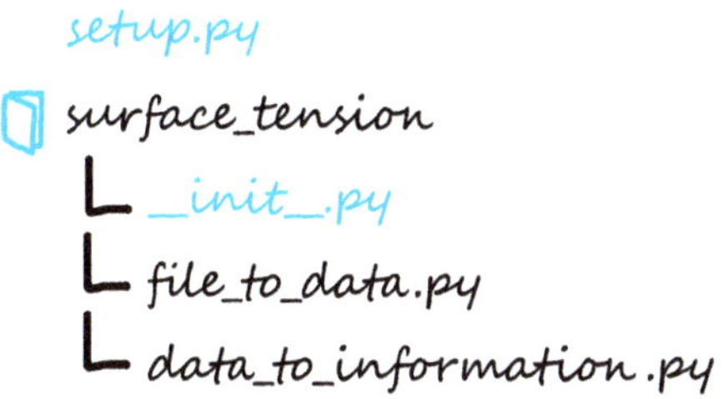

Figure A.1: Structure of files for packaging the custom package `surface_tension` containing the modules `file_to_data` and `data_to_information` via `setuptools`. The *__init__.py* file makes Python consider the *surface_tension* folder a package. The *setup.py* file serves for specifying package-specific information.

The *setup.py* file is used for providing information on the package such as name, version number, author information, keywords and a description. In our example, the file might take the form given in Listing A.1.

Listing A.1: setup.py

```
# module for packaging
from setuptools import setup

setup(
    name='surface_tension',
    version='0.1.0',
    description='pip-installable package for tasks related to \
handling surface tension data obtained from Machine X.',
    packages=['surface_tension'],
    author='Matthias Josef Hofmann',
    author_email='matthias.j.hofmann@gmail.com',
    keywords=['surface tension', 'Machine X']
    )
```

To finally create a source distribution in the project root folder, we open the command line, navigate to the folder containing the *setup.py* file and run the file with Python using the `sdist` option. The following command is to be executed in the command line:

https://doi.org/10.1515/9783110788433-011

```
python setup.py sdist
```

This will create two more subfolders in the root directory:
– *surface_tension.egg-info* and
– *dist*.

Within the *dist* folder, we find the file *surface_tension-0.1.0.tar.gz*. This is the file we were looking for. You can use it either yourself or share it with your colleagues. For installation, proceed as with any other `pip`-installable package. Explicitly: Run

```
pip install surface_tension-0.1.0.tar.gz
```

in the command line starting from the *dist* directory.

If the `pip`-installation works as expected, you will be informed via the command line. Furthermore, you can verify the installation by requesting package information of `surface_tension` via spyder's console by typing the command:

```
pip show surface_tension
```

The console yields:

```
Name: surface-tension
Version: 0.1.0
Summary: pip-installable package tasks related to handling
surface tension data obtained from Machine X.
Home-page: UNKNOWN
Author: Matthias Josef Hofmann
Author-email: matthias.j.hofmann@gmail.com
...
```

To check the new package in action, we want to read raw data from some files and print the first few lines. Please note that the functions are taken from the installed package `surface_tension` in Listing A.2.

Listing A.2: AppendixA/read_experimental_via_installed_user_package.py

```
# import custom (installed) modules with alias
import surface_tension.file_to_data as st_data
import os

# get current path
path = os.getcwd()
# info
print("script path:", path)
```

```
# script path: C:\Users\LocalAdmin\Documents\_Data_and_code\AppendixA

# build path to data
path_to_data = path + os.sep + os.pardir + os.sep + "Chapter5" +\
    os.sep + "raw_from_machine"
# info
print("data path", path_to_data)
# data path C:\Users\LocalAdmin\Documents\_Data_and_code\AppendixA\..
# \Chapter5\raw_from_machine

# loop files
for _file in os.listdir(path_to_data):
    # info
    print(_file)
    # Polysorbate20.csv

    # read file as string
    file_str = st_data.read_file_content_as_string(
        path_to_data + os.sep + _file
        )
    # get data from string
    data = st_data.get_data_from_experimental_string(
        file_str,
        show_info=False
        )
    # show top columns of resulting dataframe
    print(data.head())
    #          concentration_g_l  surface_tension_mN_m
    # 0             0.409490                 43.73
    # 1             0.147970                 43.73
    # 2             0.087851                 43.73
    # 3             0.040554                 46.13
    # 4             0.019972                 49.40

    # stop after first file
    break
```

Also for the user-defined packages, further information can be obtained via the `dir` and ? commands in the Python console within spyder.

An in-depth tutorial on how to package and publish a Python project to the Python Package Index *PyPI* (https://pypi.org/) is available online.[1]

1 https://packaging.python.org/en/latest/tutorials/packaging-projects/

B Comments to tables and columns via SQLAlchemy

Comments can be added to tables and columns upon creation. Those can be used in the sense of a *data dictionary* providing additional information on tables and columns of a database. An example showing the addition of comments to one `Table` and two `Columns` is given in Listing B.1.

Listing B.1: AppendixB\create_sqlite.py

```
# module for handling sqlite
import sqlalchemy

# create connection to on disk database "dummy.db"
engine = sqlalchemy.create_engine(
            'sqlite:///dummy.db',
            echo=True
            )

# create bound metadata
metadata = sqlalchemy.MetaData(engine)


# %% define dummy table "my_table" with comments
#

from sqlalchemy import Column

# define two table "my_table" holding columns "name" and "age"
table = sqlalchemy.Table(
    "my_table",  # table name
    metadata,  # corresponding metadata
    Column("name",
            sqlalchemy.Float,
            comment="Name of the user",  # comment on "user" column
            primary_key=True
            ),
    Column("age",
            sqlalchemy.Unicode(255),
            comment="Age of the user in years"  # column comment
            ),
    comment="A table holding information on users"  # table comment
    )

# create the table if it does not exist
table.create(checkfirst=True)


# %% get comments on table and columns
```

https://doi.org/10.1515/9783110788433-012

```
#

# information on table
print(f"Table Comment on {table.name}: {table.comment}")

# loop columns and get comments
for _c in table.columns:
    # print column name
    print(f"Column {_c}:")
    # print comment
    print(f"  - Comment: {_c.comment}")

# Table Comment on my_table: A table holding information on users
# Column my_table.name:
#   - Comment: Name of the user
# Column my_table.age:
#   - Comment: Age of the user in years
```

C A word on version control systems

Working with code is great. Working with code under a *version control system* is even better. In a somewhat oversimplified description, it allows taking snapshots of your code at certain points in time and development states in a highly defined manner. You have full control over which changed files should be included in the snapshot, and which not. The really nice thing about versioning is the option to assign a short comment to each of the snapshots and to restore each of the snapshot states at a later point in time without losing the others.

In more technical terms, version control (also known as revision control, source control or source-code management) refers to a class of systems responsible for managing changes to computer programs, documents, large web sites or other collections of information.[1] Those systems are standard in software engineering but also solutions like Dropbox™, Google Drive™ or Microsoft® OneDrive feature some kind of version control capabilities. You will be able to retrace *who* introduced *which* change/modification *when*.

In practice, relying on a version control system means that you can "fearlessly" try out some new features in your code and test them. If everything works as expected, perfect; if not, just jump back to the last "safe spot" and start again. Another benefit of version control is the ability to work on code simultaneously in a team. Each team member can work on dedicated parts of the code without interfering with a colleague's activities. After the work has been completed, only a comprehensible *merge* is required and the effort of the team members will be available overall. Of course, there is no need to work in a team to make use of a version control system. This very book was written using *Git* to its track version history. Some of the snapshots taken during the process of writing are shown in Figure C.1.

Git is a dedicated software tool for tracking changes in any set of files, usually used for coordinating efforts among programmers collaboratively working on source code during software development. Its goals include speed, data integrity and support for distributed, nonlinear workflows.[2] A good starting point for working with this kind of version control systems is—from my personal experience—*GitHub Desktop*.[3] It provides a clean interface and a visual representation of the changes made during a session without having to rely on the command line for basic tasks.

---

1 https://en.wikipedia.org/wiki/Version_control

2 https://en.wikipedia.org/wiki/Git

3 https://desktop.github.com/

https://doi.org/10.1515/9783110788433-013

Commits

Search commits   All branches

Author	Commit	Message
Matthias Hofmann	702b6c9	Conlusion ended
Matthias Hofmann	ba7c58a	Conclusion continued
Matthias Hofmann	81991cd	conclusion started
Matthias Hofmann	1223aa5	dowhy additions
Matthias Hofmann	01e32ed	work on dowhy section
Matthias Hofmann	eced56d	dowhy section continued
Matthias Hofmann	2f2fc25	causal analysis part started
Matthias Hofmann	8ad4a44	further captions cosmetics
Matthias Hofmann	d95268f	Appendix user package
Matthias Hofmann	1deacbf	figure caption cosmetics chapters 4 and 5
Matthias Hofmann	10163e1	Cover page options
Matthias Hofmann	ba18507	two spyder screenshots added
Matthias Hofmann	e8674f4	marginnotes removed

Figure C.1: Screenshot of some version control snapshots (also referred to as *commits*) taken in the course of writing this book using *Git*.

D Overview of utilized Python and package versions

The versions of Python and its packages listed in Table D.1 were used for developing and running the code snippets shown herein.

Table D.1: List of the packages used herein with the corresponding version numbers.

Module	Version
Python	3.8.5
dowhy	0.7
numpy	1.19.2
pandas	1.1.3
seaborn	0.11.0
setuptools	50.3.1.post20201107
SQLAlchemy	1.3.20
sqlalchemy-migrate	0.13.0
surface-tension	0.1.0

Please note that, after installation via `pip`, our custom-built package `surface-tension` also appears in the list. If you want to compare these versions with your installation, just run `pip list` in the console of the spyder integrated development environment (IDE). This will print a summary of the installed packages alongside its versions.

https://doi.org/10.1515/9783110788433-014

E Extracting data via pd.read_csv

As an alternative to the previously introduced *data* and *information* extraction from an experimental raw data file relying on regular expressions, the code shown in Listing E.1 provides a possible solution based on the pandas-function read_csv. Certainly, there may be other (and possibly more elegant) ways out there. The proposed process for reading the *data* part includes the following steps:

- reading the contents of the entire file as the index column of the obtained DataFrame,
- selecting the *data* part between the characteristic file parts === *COLLECTED DATA* === and === *SUMMARY INFORMATION* ===,
- interpreting the index column as raw column,
- separating concentration and surface tension for the raw column using *list comprehensions*, and
- applying some cosmetics such as using the column names suggested from the experimental raw data file and a type conversion of the numerical values.

The procedure for extracting the *information* part is quite similar. The major difference is to consider only the range of the file before the === *COLLECTED DATA* === separator.

Listing E.1: AppendixE\read_file_via_pd_read_csv.py

```
# import module for reading data
import pandas as pd

# specify file
file = "Polysorbate40.csv"

# read
raw = pd.read_csv(
            file,
            sep="\n",  # new line separator as column separator
            index_col=0  # use all string as column
            )
# info
print(raw)


# %% get "data" part via index range
#

data = raw.loc[
        "=== COLLECTED DATA ===":
        "=== SUMMARY INFORMATION ===",  # select range of rows
        :  # select all columns
```

https://doi.org/10.1515/9783110788433-015

```
        ]
# info
print(data)
# Empty DataFrame
# Columns: []
# Index: [=== COLLECTED DATA ===, Concentration / [g/L],Surface
# tension / [mN/m], 4.0790e-01,43.78, 1.4911e-01,43.84,
# 8.8993e-02,43.78, 3.9720e-02,46.68,
# 1.9793e-02,49.43, 9.2199e-03,53.69,
# 2.7936e-03,57.40, === SUMMARY INFORMATION ===]

# use index as column as value column
data = data.reset_index(level=0)
# set name of previous index column to "raw"
data.columns = ["raw"]

# discard first and last row
data = data.iloc[1:-1,:]

# separate columns via list comprehensions; split at ","
data["column_1"] = [i[0] for i in data["raw"].str.split(",")]
data["column_2"] = [i[1] for i in data["raw"].str.split(",")]

# discard "raw" column
data = data.drop(columns=["raw"])

# use first line as header
data.columns = data.iloc[0,:]

# use second to last row as data part
data = data.iloc[1:,:]

# convert column to float
for _c in data.columns:
    data[_c] = data[_c].astype(float)

# show results
print(data)
# 1  Concentration / [g/L]  Surface tension / [mN/m]
# 2               0.407900                     43.78
# 3               0.149110                     43.84
# 4               0.088993                     43.78
# 5               0.039720                     46.68
# 6               0.019793                     49.43
# 7               0.009220                     53.69
# 8               0.002794                     57.40


# %% get "information" part via index range
```

```
#

information = raw.loc[
        :"=== COLLECTED DATA ===",  # select range of rows
        :  # select all columns
        ]

# use index as column
information = information.reset_index(level=0)
# set name of previous index column to "raw"
information.columns = ["raw"]

# discard last row
information = information.iloc[:-1,:]

# separate columns via list comprehensions; split at ": "
information["column_1"] = [i[0].strip() for i in
                                information["raw"].str.split(": ")]
information["column_2"] = [i[1].strip() for i in
                                information["raw"].str.split(": ")]

# discard "raw" column
information = information.drop(columns=["raw"])

# transpose results
information = information.T

# use first line as header
information.columns = information.iloc[0,:]
# use second to last row as data part
information = information.iloc[1:,:]

# reset index
information = information.reset_index(drop=True)

# show results
print(information.T)
#                                         0
# column_1
# Sample                       Polysorbate40
# Measurement performed on  2022-01-28 13:37:44
# Operator                     Matthias Hofmann
# Device                        Fancy Machine X
# Device ID                           Y0139836
```

Go through the previous code snippet step by step to understand the process in detail. What are the advantages and disadvantages of this approach compared to using regular expressions applied to the file content read as a `string`?

F Installation of an ODBC driver

Installing an SQLite-Open Database Connectivity (ODBC) driver enables connecting our exemplary database *surface_tension.db* to Microsoft® Power BI Desktop®. The driver can be downloaded, for instance, from:

- http://www.ch-werner.de/sqliteodbc/, or
- https://www.devart.com/odbc/sqlite/.

After running the installation wizard, open the *ODBC Data Source Administrator* and "Add" your new *User Data Source*. In this procedure, an appropriate Data Source Name (DSN) has to be defined.

https://doi.org/10.1515/9783110788433-016

List of Figures

https://doi.org/10.1515/9783110788433-017

Bibliography

[1] M. Bayer, SQLAlchemy. In: The Architecture of Open Source Applications, A. Brown and G. Wilson (eds.), Series The architecture of open source applications, vol. 2, AOSA, 2012.

[2] D. Kahneman, Thinking, fast and slow. In: Psychology/economics, 1 paperback ed. Farrar Straus and Giroux, New York, 2013.

[3] J. Pearl and D. Mackenzie, The book of why: The new science of cause and effect, 1st ed. Basic Books, New York, 2018.

[4] S. Raschka and V. Mirjalili, Python machine learning: Machine learning and deep learning with Python, scikit-learn, and TensorFlow. Expert insight. Packt Publishing, Birmingham and Mumbai, second edition, fourth release, [fully revised and updated] edition.

[5] A. Sharma and E. Kiciman, DoWhy: An End-to-End Library for Causal Inference.

[6] M. Waskom, seaborn: Statistical data visualization. J. Open Sour. Softw. 6(60):3021, (2021).

[7] M.D. Wilkinson, M. Dumontier, I. Jsbrand Jan Aalbersberg, G. Appleton, M. Axton, A. Baak, N. Blomberg, J.-W. Boiten, L. Bonino da Silva Santos, P.E. Bourne, J. Bouwman, A.J. Brookes, T. Clark, M. Crosas, I. Dillo, O. Dumon, S. Edmunds, C.T. Evelo, R. Finkers, A. Gonzalez-Beltran, A.J.G. Gray, P. Groth, C. Goble, J.S. Grethe, J. Heringa, P.A.C. 't Hoen, R. Hooft, T. Kuhn, R. Kok, J. Kok, S.J. Lusher, M.E. Martone, A. Mons, A.L. Packer, B. Persson, P. Rocca-Serra, M. Roos, R. van Schaik, S.-A. Sansone, E. Schultes, T. Sengstag, T. Slater, G. Strawn, M.A. Swertz, M. Thompson, J. van der Lei, E. van Mulligen, J. Velterop, A. Waagmeester, P. Wittenburg, K. Wolstencroft, J. Zhao and B. Mons, The FAIR Guiding Principles for scientific data management and stewardship. Sci. Data 3:160018, (2016).

[8] N. Yau, Data points: Visualization that means something, 1. aufl. edition. Wiley, Indianapolis, 2013.

https://doi.org/10.1515/9783110788433-018

Index

https://doi.org/10.1515/9783110788433-019

www.ingramcontent.com/pod-product-compliance
Lightning Source LLC
Chambersburg PA
CBHW061418210326
41598CB00035B/6253